SpringerBriefs in Law

SpringerBriefs present concise summaries of cutting-edge research and practical applications across a wide spectrum of fields. Featuring compact volumes of 50 to 125 pages, the series covers a range of content from professional to academic. Typical topics might include: • A timely report of state-of-the art analytical techniques • A bridge between new research results, as published in journal articles, and a contextual literature review • A snapshot of a hot or emerging topic • A presentation of core concepts that students must understand in order to make independent contributions SpringerBriefs in Law showcase emerging theory, empirical research, and practical application in Law from a global author community. SpringerBriefs are characterized by fast, global electronic dissemination, standard publishing contracts, standardized manuscript preparation and formatting guidelines, and expedited production schedules

More information about this series at http://www.springer.com/series/10164

Romola Adeola

Emerging Issues in Internal Displacement in Africa

 Springer

Romola Adeola
Centre for Human Rights
University of Pretoria
Pretoria, South Africa

ISSN 2192-855X   ISSN 2192-8568   (electronic)
SpringerBriefs in Law
ISBN 978-3-030-64561-8   ISBN 978-3-030-64562-5   (eBook)
https://doi.org/10.1007/978-3-030-64562-5

This Springer imprint is published by the registered company Springer Nature Switzerland AG
The registered company address is: Gewerbestrasse 11, 6330 Cham, Switzerland

Contents

1   Introduction: Emerging Issues, Internal Displacement
    and Regional Legal Protection in Africa.   1
    References.   4

2   Emerging Issues: Conceptual Clarification   5
    2.1   Climate Change   5
    2.2   Technology.   8
    2.3   Xenophobia   10
    2.4   Harmful Practices.   11
    2.5   Generalized Violence   12
    2.6   Development Projects   13
    References.   13

3   Emerging Issues and Internal Displacement in Africa:
    Interlinkages   19
    3.1   Climate Change   19
    3.2   Technology.   22
    3.3   Xenophobia   24
    3.4   Harmful Practices.   27
    3.5   Generalised Violence   29
    3.6   Development Projects   31
    3.7   Conclusion   32
    References.   34

4   Legal Protection in Emerging Contexts   41
    4.1   Process.   41
          4.1.1   UN Guiding Principles on Internal Displacement   42
          4.1.2   AU Convention on Internally Displaced Persons.   44
          4.1.3   ICGLR Internally Displaced Persons Protocol   47
    4.2   Content.   49
          4.2.1   The Right Not to Be Arbitrarily Displaced   52

     4.3   Conclusion   59
     References.   60
5   **Conclusion and Recommendations**   63
     5.1   Conclusion   63
     5.2   Recommendations   66
     5.3   Implication for Refugees and Migration Studies.   67
     References.   69

Index.   71

Chapter 1
Introduction: Emerging Issues, Internal Displacement and Regional Legal Protection in Africa

Abstract Internally displaced persons are persons who remain within the border of the state of displacement. Unlike refugees, these persons do not cross an international border. Over the last decades, the issue of internal displacement has become a prevalent concern. While scholarly attention to this issue has mostly centred around conflict, there are emerging root causes of internal displacement in Africa for which significant research is required. This book focuses on these emerging root causes of internal displacement. This chapter introduces the discussion in this book.

Keywords Internally displaced persons · Internal displacement · African Union · Africa

The issue of internal displacement has emerged as a global concern. As a pertinent issue requiring sustainable solutions, calls for concrete actions have emerged and normative guidance has begun to surface at various governance levels. Besides global estimate on forcibly displaced populations which accentuate the fact that most forcibly displaced persons are within the borders of the state, the multifaceted nature of the root causes of this form of displacement coupled with the protection gaps associated with this displacement raises the imperative for sustainable solutions. However, much of the research on the issue of internal displacement in Africa has focused on conflict which displaced at least 16.8 million people in 2018.[1] There are good reasons for this, principal among which is that it is the primary driver of internal displacement in Africa.

However, there are other emerging root causes of internal displacement, for which significant attention is required, globally and within regional contexts including Africa. The central aim of this book is to examine these emerging root causes of internal displacement that have not received considerable attention in the narrative on internal displacement within the African regional context. This book engages the

---

[1] Internal Displacement Monitoring Centre (2019, 8).

© The Author(s), under exclusive license to Springer Nature Switzerland AG 2021
R. Adeola, *Emerging Issues in Internal Displacement in Africa*, SpringerBriefs in Law, https://doi.org/10.1007/978-3-030-64562-5_1

1

subject of internal displacement leveraging on the regional norm on internal displacement, the African Union (AU) *Convention for the Protection and Assistance of Internally Displaced Persons in Africa* (Kampala Convention) adopted by the AU Heads of State and Government in October 2009.[2]

Over the last decades, significant research has emerged in the internal displacement context considering the imperative for a normative framework in response to the challenge. Early research was advanced through the contribution of notable think-tanks, particularly, the Brookings-LSE Project on Internal Displacement in Africa and the Brookings-SAIS Project.

Following the adoption of the Kampala Convention, research in the field of internal displacement in Africa has notably flourished in various areas from gender perspectives to the accountability of non-state actors and from the development of the regional treaty to its impact within various contexts. While there has also been significant research on internal displacement from arts, health, social sciences and humanities, an evident gap in the literature relates to research on the emerging areas discussed in this book. While this book provides groundwork necessary for taking these issues forward, it further reflects on the provisions of the regional normative framework in the furtherance of protection for IDPs within the emerging contexts.

This research applies a legal analytical approach in discussing the legal protection of IDPs in the context of emerging root causes of internal displacement in Africa. Given the dearth of knowledge on these emerging areas, this book further adopts a broader exploratory social science method drawing on primary and secondary sources. The pertinence of this approach resonates from the fact that it provides an optic for considering issues for which little knowledge exists. This method is employed in 'clarifying concepts, gathering explanations, gaining insights, refining problems and ideas, and forming hypotheses.'[3] It gives a viewpoint from which further research may be developed. This book utilises this method to provide insights into these emerging areas to tease out the connections inherent in these issues and internal displacement for which significant attention is required in addressing the various dimensions of internal displacement. In this regard, this research provides groundwork for the development of future research in the area of internal displacement both in Africa and beyond.

A plethora of rich scholarly research have emerged on the imperative of law in internal displacement. This research has reinforced the salience of law and policy formation in the protection of IDPs more generally and the pertinence of legal guidance on various causes of internal displacement. While global structures are imperative to this process, the role of regional structures have become of importance.[4] Regional structures are pertinent forums for negotiating broad-based solutions on

---

[2] Kampala Convention (2009).

[3] Andrew et al. (2020, 9). See also Blaikie (2007, 70), Henn et al. (2009, 134), McNabb (2010, 96), Sarantakos (2013, 150–151).

[4] In international law, the benefit of regionalism as a law-making technique has been firmly established. See Mittelman (1999, pp. 25–53), Agnew (2013, p. 130), Bach (2014, p. 181), Matthew and Harley (2016, p. 60).

seemingly difficult issues, due to the fact that these structures leverage on shared values, proximity, collective loyalties and commonality of issues.[5]

In Africa, the narrative upon which the Organisation of African Unity (OAU) emerged was essentially political – with the objectives of decolonisation, anti-imperialism and freedom for the African peoples. [6] With the end of the Cold War and decolonisation of Africa, regionalism ('new regionalism') among states became a method for promoting mutual cooperation in a plethora of areas including economic, social, political and cultural facets.[7] Through regionalism, and inter-regionalism (which entails cooperation between regions), issues such as trade, development, economic integration, human rights and democratisation have been fostered. The idea of regionalism has become a conduit for addressing challenges based on mutual interests and the commonality of purpose. This resonates with the motivation behind the development of the regional norms on forced migration. As these norms afford a premise on which to develop a common guidance, an understanding of these norms and the extent to which they provide legal protection is imperative in shaping the regional landscape. In Africa, the corpus on internal displacement draws on texts at global, regional and sub-regional levels. Principal among these, however, is the Kampala Convention. Drawing on this regional framework, this book leverages on the premise that an understanding of these norms is useful in defining the extent of legal protection, gaps that need to be addressed and areas where engagement with states and non-state actors ought to be amplified for the adequate protection of IDPs to further the global agenda towards finding sustainable solutions to the issue of internal displacement.

This book charts new territory in the field of internal displacement. It examines emerging root causes of internal displacement for which there is a need for significant research. With a focus on Africa, this book draws on evidence of intersections between six emerging issues and internal displacement, arguing that for a

---

[5] The International Law Commission (ILC) details the significance of regionalism in observing that it offers a 'privileged forum for international lawmaking.' Regionalism leverages on the geographical interest, international solidarity and the concept of brotherhood which often creates an impetus for compliance. The ILC further observed that '[i]t is often assumed that international law is or should be developed in a regional context because the relative homogeneity of the interests or outlooks of actors will then ensure a more efficient or equitable implementation of the relevant norms. The presence of a thick cultural community better ensures the legitimacy of the regulations and that they are understood and applied in a coherent way. This is probably the reason for why human rights regimes and free trade regimes have always been commenced in a regional context - despite the universalist claims of ideas about human rights or commodity markets.' International Law Commission (2006, para. 205).

[6] See *Charter of the Organisation of African Unity*, adopted on 25 May 1963; Asante (1997, p. 34), Francis (2006, p. 24), Hartmann (2016, p. 273), Bach (2016, p. 78).

[7] New regionalism, unlike the old, became the means through which states strengthened mutual cooperation on areas of global priorities. Paragraph 6 of the Constitutive Act of the African Union (AU) reflects this in emphasising the collective desire of AU Member States to 'take up the multi-faceted challenges that confront our continent and peoples in the light of the social, economic and political changes taking place in the world'. See *Constitutive Act of the African Union* adopted on 11 July 2000, para 6 of Preamble.

comprehensive approach to addressing the root causes of internal displacement there is a need to tackle these issues. These root causes are: climate change, technology, xenophobia, harmful practices, generalized violence and development projects.

References

African Union Constitutive Act of the African Union, adopted on 11 July 2000

African Union convention for the protection and assistance of internally displaced persons in Africa, adopted on 23 Oct 2009

Agnew J (2013) The new regionalism and the politics of the regional question. In: Loughlin J, Kincaid J, Swenden W (eds) Routledge handbook of regionalism and federalism. Routledge, Abingdon

Andrew DPS, Pedersen PM, McEvoy CD (2020) Research methods and design in sport management. Human Kinetics, Champaign

Asante SKB (1997) Regionalism and Africa's development: expectations, reality and challenges. Macmillan Press, London

Bach DC (2014) Regionalism in Africa: concepts and context. In: Hentz JJ (ed) Routledge handbook of African security. Routledge, Abingdon

Bach DC (2016) Regionalism in Africa: genealogies, institutions and trans-state networks. Routledge, Abingdon

Blaikie N (2007) Approaches to social enquiry: advancing knowledge. Polity Press, Cambridge

Charter of the Organisation of African Unity, adopted on 25 May 1963

Francis DJ (2006) Uniting Africa: building regional peace and security systems. Ashgate, Farnham

Hartmann C (2016) Sub-Saharan Africa. In: Börzel TA, Risse T (eds) The Oxford handbook of comparative regionalism. Oxford University Press, Oxford

Henn M, Weinstein M, Foard N (2009) A critical introduction to social research. Sage, Thousand Oaks

Internal Displacement Monitoring Centre (2019) Africa report on internal displacement. Internal Displacement Monitoring Centre, Geneva

International Law Commission, fifty-eighth session, Fragmentation of international law: difficulties arising from the diversification and expansion of international law: report of the study group of the International Law Commission finalized by Martti Koskenniemi UN Doc A/CN.4/L.682, 13 Apr 2006

Matthew P, Harley T (2016) Refugees, regionalism and responsibility. Edward Elgar Publishing, Cheltenham

McNabb DE (2010) Research methods for political science. M.E. Sharpe, New York

Mittelman JH (1999) Rethinking the new regionalism in the context of globalization. In: Hettne B, Inotai A, Sunkel O (eds) Globalism and the new regionalism. Macmillan Press, London

Sarantakos S (2013) Social research. Macmillan Press, London

Chapter 2
Emerging Issues: Conceptual Clarification

Abstract Having a significant understanding of the various root causes discussed in this book is important in drawing the nexus between these root causes and internal displacement. There are six emerging root causes of internal displacement discussed in this book, namely, climate change, technology, xenophobia, harmful practices, generalized violence and development projects. This chapter discusses these root causes providing an insight into these issues.

Keywords Internal displacement · Climate change · Technology · Xenophobia · Harmful practices · Generalized violence · Development projects

2.1 Climate Change

That climate change is happening is hardly debatable. The increase in global average surface temperature, the footprints of anthropogenic forcing, the extensive body of knowledge on ocean acidification, torrential rain patterns, melting ice caps, increased sea level rise and severe heatwaves lend credence to the global clarion call on addressing the causes of climate change, mitigating its impacts and adapting to its consequences.[1] Although empirical evidence on the variability in global surface temperature is detailed and overwhelming,[2] some have described climate change as 'a hoax' and a false narrative to impede developmental gains.[3] However, the science reveals differently and affirms that human activities is significantly driving the

---

[1] See Hoegh-Guldberg (2012, 21), Pouch and Zaborska (2015, 75).

[2] Notably the IPCC has contributed to knowledge formations on the fact that climate change is happening and its increasing impacts. For detailed reports of the IPCC since its formation in 1988. In understanding this role, see Kimuyu (2016), UN Human Rights Council, *Report of the Office of the United Nations High Commissioner for Human Rights on the Relationship between Climate Change and Human Rights*, Annual Report of the United Nations High Commissioner for Human Rights and Reports of the Office of the High Commissioner and the Secretary-General, UN Doc A/HRC/10/61 (15 January 2009), 4.

[3] See generally Leroux (2006), Michaels and Balling (2009), Booker (2009), Bell (2011), Amber (2011), Burnett (2014), Darwall (2019).

change with an overconcentration of greenhouse gases in the atmosphere. According to the Intergovernmental Panel on Climate Change (IPCC), '[h]uman influence on climate has been the dominant cause of observed warming since the mid-20th century'.[4] Human activities had warmed the world by 0.87 °C (±0.12 °C) between 2006 and 2015 in comparison to the pre-industrial period (1850 and 1900).[5] If current trends are not reversed, 'the world would reach human induced global warming of 1.5 °C around 2040.'[6] In 2014, the IPCC noted that CO2 attributable human activities was responsible for 78% of the total greenhouse gases in the atmosphere.[7] According to the United States Fourth National Climate Assessment, 'the world's oceans have absorbed about 93% of the excess heat caused by greenhouse gas warming since the mid-20th century, making them warmer and altering global and regional climate feedbacks'.[8] Scientists warn that without restraint in fossil fuel usage, how the earth will respond may likely be without geological precedence.[9] While the situation of small island developing states (SIDS) such as Tuvalu, the Maldives and Kiribati has been instrumental in shaping the ethical optics,[10] increased droughts, torrential rainfalls and famine in the eastern region of Africa equally makes this point.[11]

With the risk of the Earth becoming a 'hothouse state' if current trends in climate change are not addressed,[12] the past four decades has significantly witnessed discussions on the multi-dimensional impacts of climate change. These impacts have garnered interests from a plethora of fields and as the world increasingly moves towards the Anthropocene Era (defined by an increase in human footprints on the geological conditions of the earth),[13] climate change has become an issue of global governance.[14] Given the multifaceted nature of its impact, its convergence with other topical issues has brought it into spotlight with these issue in parallel conversations. An

---

[4] Earlier, in its fifth assessment report, the IPCC emphasised high plausibility that 'more than half of the observed increase in global average surface temperature from 1951 to 2010 was caused by the anthropogenic increase in greenhouse gas concentrations and other anthropogenic forcings together.' Intergovernmental Panel on Climate Change (2013, 16), Intergovernmental Panel on Climate Change (2018, 53).

[5] Intergovernmental Panel on Climate Change (2018, 81); see Schurer et al. (2017, 563–567).

[6] Intergovernmental Panel on Climate Change (2018, 81).

[7] Intergovernmental Panel on Climate Change (2014).

[8] United States Global Change Research Program (2017).

[9] Mathez (2009), Foster et al. (2017, 14845).

[10] See generally Burns (2000), Climate Change Secretariat (UNFCCC) (2005), Barnett and Campbell (2010), Munro (2010, 145).

[11] Ddamulira (2016, 257–262), Egeru et al. (2016, 69), Daily Nation (2018).

[12] See McGrath (2018), Watts (2018), Poole (2018), Chestney (2018), Zalasiewicz et al. (2018), Gabbatiss (2018).

[13] Crutzen and Steffen (2003, 251–257), Ribot (2014, 667–705), Hoffman et al. (2015, 3), Stager (2016, 17), Delanty and Mota (2017, 20(1) 9–38), Wright et al. (2018, 455–471).

[14] See Ackerly and Vandenbergh (2008, 553), Kirton and Kokotsis (2016).

area of such parallel conversation within the focus of this book relates to the issue of population movements, specifically, internal displacement.

Since the report of the Intergovernmental Panel on Climate Change (IPCC) linking migration to climate change as early as 1990,[15] the debate on the linkage between climate change and mobility has flourished. There are essentially two schools of thought that have emerged on the linkage between the subject – the maximalists and minimalists.[16] Maximalists, often regarded as alarmists, consider climate change as directly linked to significant population movements. This argument, often canvassed by environmental activists, civil society and the media has drawn significant attention to the growing impact of climate change, significantly on the issue of mobility. But given the existence of multi-causality in the movement patterns associated with climate change, maximalists are often criticised for not being able to substantially reflect the multi-causal linkages. Yet, their arguments on the causal connection between climate change and mobility is not without substance in view of the potential impacts of climate variability and empirical studies that are emerging on the impact of climate change on population movements in regions such as the Asia-Pacific and East Africa. Minimalists, or sceptics, on the other hand, are driven by a more evidence-based approach to the narrative. These group often raise inquiry into the modalities of arriving at estimates of population movements due to climate change. To sceptics, the fact that climate change is happening is not in doubt. However, to assert that there is a causal link between climate change and mobility ought to be approached with caution. The lack of extensive empirical evidence to support the mono-causal argument evident in the prevalent sensational outlook of the discourse is the prime basis for their restraint.

But despite the divergence in persuasions, there is a consensus between both minimalists and maximalists that the discourse on climate change in the context of mobility presents an opportunity to curate sustainable solutions to a possible dimension of climate change and the vulnerability of populations on the move within this context. Moreover, the normative discourse on climate change and population movement has, in more recent years, gained global validation through studies, UN resolutions and international soft norms including the *Global Compact on Safe, Orderly and Regular Migration*.[17] The relevance of the law in the global governance context is increasingly of interest given the need for normative protection and the imperative of closing existing gaps in solving pertinent global challenges.

---

[15] According to the IPCC, 'the gravest effects of climate change may be those on human migration as millions are displaced by shoreline erosion, coastal flooding and severe drought.' See Intergovernmental Panel on Climate Change (1990, 5–9), Intergovernmental Panel on Climate Change (1992, 103).

[16] See Myers (1993, 752–761), Suhrke (1993), Boykoff and Roberts (2007), Johnson (2010), Martin (2010, 397), Mayer (2011, 91), Jakobeit and Methmann (2012, 301–314), Piguet (2013, 148–162), Campbell and Bedford (2014, 177), Ferris (2014), Thornton (2018, 14–15).

[17] *Global Compact on Safe, Orderly and Regular Migration, adopted by the Intergovernmental Conference on the Global Compact for Migration* 2018.

Over the last decade, the global lexicon on climate mobility has flourished. In early discussions, the notion of 'environmental refugees' was used to curate knowledge.[18] This notion was used to refer to persons 'who have been forced to leave their traditional habitat, temporarily or permanently, because of a marked environmental disruption (natural and/or triggered by people) that jeopardised their existence and/ or seriously affected the quality of their life.'[19] Drawing on situations in the Caribbean, sub-Saharan Africa and Asia-Pacific, Myers projected that about '200 million people' could become environmental refugees.[20] As discussions have progressed, other terminologies have begun to emerge including 'environmental migrants', 'climate migrants', 'environmental emigrants', 'climate refugees', 'climate displaced persons' and 'environmentally displaced persons'. However, the modalities for classification are not always clear.

This chapter draws on the tri-dimensional categorisation on the climate-mobility nexus from the Conference of State Parties to the United Nations Framework Convention on Climate Change (UNFCCC).[21] At the Conference, the issue of displacement is recognised as one of the climate-mobility dimensions. While significant discussion has begun to emerge on cross-border displacement, not much exists on the nexus between climate change and internal displacement. With specific reference to Africa, this research examines climate change within the context of internal displacement.

2.2 Technology

From the early periods of human evolution, the epochs of cognition around technology can be broadly subdivided into the agrarian and industrial revolutions.[22] The first phase of the agrarian revolution was essentially around the use of enhanced agricultural cognition in many parts of the world and spanned from the Neolithic era in 10,000 BC to the Green era in the 1960s. In the Neolithic era, agriculture was centred on 'systems of pastoral animal breeding and systems of slash-and-burn cultivation.'[23] Archaeological evidence in the Fertile Crescent – northern areas of the Nile, eastern Mediterranean and Iraq – demonstrates the existence of human cognition towards planting seeds and harvesting them.[24] There were developments

---

[18] See El Hinnawi (1985), Jacobson (1988, 465–477).

[19] El Hinnawi 1985, 4.

[20] Myers (1997, 168), Myers (2002, 609).

[21] Decision 1/CP.16: The Cancun Agreements: Outcome of the work of the Ad Hoc Working Group on Long-term Cooperative Action under the Convention, adopted at the 16th Session of the United Nations Framework Convention on Climate Change Conference of the Parties, Cancun, Mexico (29 November – 10 December 2010), para 14(f).

[22] See Schwab (2017).

[23] Mazoyer and Roudart (2006, 4).

[24] Meyer (2017).

in these periods on methods of agricultural production. For instance, the movement of the Bantu speakers towards the Great Lakes in Africa in the period around 1000 BC was premised on enhanced agrarian practices and the emergence of iron-working technology. From the sixteenth century when the agricultural revolution occurred in England,[25] there emerged significant transformations in agrarian methods and techniques which, mostly resulted from the need to expand markets. By the mid-seventeenth century, there were significant cropping innovations. In this period innovations 'led to increases in grain output per acre, and a rise in total output evidenced by rising grain exports.'[26] Around this period, the industrial revolution emerged. Within this revolution are four phases that can be discerned.

The first industrial revolution, which began the process that has led to the creation of 'the affluent societies of today'[27] emerged in the mid-eighteenth century. This period saw the emergence of hand to machine forms of production: such as the creation of seed drills and iron ploughs. One of the prominent areas in which there was significant development was the textile industry. There were also improvements in agricultural production processes with irrigation facilities and high-yielding varieties of agricultural crops.

The second industrial revolution, which started around 1870, began the process of electrification.[28] Building on the first industrial revolution, this period saw the emergence of new technologies that leveraged on electricity and telecommunications. There were notable developments in various fields.[29] The growth in research in the area of materials science led to enhanced production of steel. An improved understanding of thermodynamics also propelled notable innovations in various industries. In this period, the automobile industry emerged with Karl Benz's invention being the first patented automobile in 1886. Also, the Wright Brothers achieved an important milestone by creating the first controlled airplane.

The third industrial revolution, essentially the age of internet, has significantly witnessed massive digitalisation of manufacturing. In this age of industrialisation, economic paradigms have leveraged on the potentials of digital technology both in transforming the means of production and in the global chain supply. The third industrial revolution has also influenced various aspects of societies ranging from commerce to culture. Within this revolution has also emerged the rapid movement of goods and services with societies becoming increasingly multi-local and ideas travelling in real-time and shaping global narratives in economics, trade, entertainment and increasingly, politics. The lateral force of this revolution has increasingly made technology a part of human development.

With the emergence of artificial intelligence, automation and big data, characterised by expanding digital infrastructure, the presence of a fourth industrial

---

[25] See Kerridge (1967); Overton (1989, 9).

[26] Overton (1996, 5).

[27] Deane (1979, 4).

[28] Smith and Free 2016.

[29] Mokyr (n.d.).

revolution has become a global rhetoric.[30] Increasingly, there is a presence of a fourth phase 'characterized by a much more ubiquitous and mobile internet, by smaller and more powerful sensors that have become cheaper, and by artificial intelligence and machine learning'.[31] There is an increasing level of sophistication building around how we conceive of digital technology that has enormous potentials both for human cognition and societal advancements. Increasingly, the world is experiencing a diffusion of ideas with significant implications on the global political economy, socio-economic welfare of populations and national economic paradigms. However, with this transmutation of society has also emerged significant challenges.

Research on the impact of technology has advanced in two directions at once – exploring both negative and positive impacts of technology and what this means for the global narrative on political power, economic control, human rights, international peace and security. On the negative impacts, the mass surveillance of populations, the deployment of lethal autonomous weapons in warfare and computational propaganda,[32] among others, has raised significant concerns particularly with respect to a plethora of rights including the rights to life, privacy, physical integrity, non-discrimination and expression. While much of these discussions have centred on human rights violations, the nexus between technology and internal displacement is yet to be discussed. With the advent of the fourth industrial revolution and the emergence of artificial intelligence, there is a need to pay attention to the linkage between technology and internal displacement. Hence, the gap which this book addresses.

2.3   Xenophobia

In many parts of the world, the issue of xenophobia has emerged as a concern.[33] The etymology of the word connotes fear (*xenos*) of the foreign (*phobos*). While the fear of the foreign relates broadly to a plethora of contexts that may be classified as foreign on which prejudice may derive, the term 'xenophobia' is typically contextualized with reference to prejudices against non-nationals of a particular state. The increasing tide of populist rhetoric and its implication for nation-state ideological formations has driven the narrative on xenophobia and legitimated its popularization in far-right movements, political spaces as with national and local dialogues and ideation platforms on identities.

---

[30] Schwab and Davis (2018), Skilton and Hovsepian (2018), Penprase (2018).

[31] Rifkin (2012).

[32] See Scharre (2018), Gorwa (2019, 86), Snowden (2019).

[33] Crush and Pendleton (2004), Rydgren (2004, 123), Lewin-Epstein and Levanon (2005, 90), Roemer and van der Straeten (2006, 251), Berezin (2006, 273), Nyamnjoh (2006), Saideman and Ayres (2008), Yakushko (2009, 36), Neocosmos (2010), Bordeau (2010), Misago et al. (2015), Farmbry (2019).

At the heart of the narrative is the notion of dislike for what is considered foreign or the 'rejection of outsiders' which may manifest in various dimensions. This rejection of the foreigner as the 'foreign-other' is cast against the backdrop of para-doxes, the ideation of free movement of goods and services and the restriction on producers of such goods and services. In many parts of the world, the issue of xeno-phobia is becoming an increasing determinant of who has access to social forma-tions and what the nature of access to the nation-state and its societal benefits are. As with racism, xenophobia is predicated on bias, however, on the ideation of nationality and citizenship. It is pertinent to note that xenophobia may be expressed through a plethora of forms, including physical forms such as prejudicial behav-iours and outright attacks; through structural forms such as policies that criminal-izes migration and through virtual forms: such as social media and online platforms. While much of the attention on the subject in the context of mobility has congre-gated around situations of mass expulsion,[34] the nexus between xenophobia and internal displacement has not gained significant attention.

2.4 Harmful Practices

Rooted in traditions, customs and belief systems, harmful practices are prevalent concerns in many societies. These practices relate to behavioural patterns, attitudi-nal orientations and practices that may emerge from conducts, accepted societal norms, cultural traditions or mannerisms and constitute violations of human rights. These behaviours, attitudes and practices cut across a wide range of issues. Harmful practices are a prevalent form of violence, the implication of which are long lasting in many societies and on a plethora of groups. Not only do they compound vulner-abilities in many instances, they reinforce rhetoric of marginalisation, deprivation and powerlessness in various contexts. Moreover, these practices are perpetuations of discrimination,[35] the effect of which is detrimental to the furtherance of sustain-able development. They also reinforce inequalities and in certain instances, have serious impacts on mental and physical health. While it is important to understand

---

[34] Henckaerts (1995), Bekker (2009, 262), Adeola (2015, 253).

[35] In General Recommendation 31, the Committee on the Elimination of Discrimination against Women (CEDAW) and the Committee on the Rights of the Child observed that '[h]armful prac-tices are persistent practices and forms of behaviour that are grounded in discrimination on the basis of, among other things, sex, gender and age, in addition to multiple and/or intersecting forms of discrimination that often involve violence and cause physical and/or psychological harm or suf-fering. The harm that such practices cause to the victims surpasses the immediate physical and mental consequences and often has the purpose or effect of impairing the recognition, enjoyment and exercise of the human rights and fundamental freedoms of women and children.' See UN Committee on the Elimination of Discrimination against Women *Joint general recommendation no. 31 of the Committee on the Elimination of Discrimination against Women/general comment no. 18 of the Committee on the Rights of the Child on harmful practice*, UN Doc CEDAW/C/GC/31-CRC/C/GC/18 (14 November 2014), para 15.

factors that perpetuate them, it is equally imperative to understand their impacts. Paying attention to this practice and its nexus to internal displacement is key in developing normative and institutional response that can adequately address protect IDPs within this context. This is important as regular law and institutional systems may not have adequate provisions or competences given the peculiarity of such practices and the context within which internal displacement occurs.

2.5 Generalized Violence

Situations of generalized violence are instances of widespread terror or indiscriminate attack against populations within a part of a state. While such situations of violence may resonate in the context of armed conflict, the notion of generalized violence extends beyond the parameters of conventional armed conflict. It expansively includes violence that may fall below the threshold of armed conflict as explicated under international humanitarian law. These forms of attack are often geared towards instilling fear. They may be carried out by state or non-state actors in given instances. While this root cause of internal displacement is quite prevalent, the discussion on its dimensions in the literature are few and far between. Before its inclusion in the Guiding Principles on Internal Displacement, generalized violence emerged within the regional normative framework on refugees in the Latin American region – the Cartagena Declaration.[36] The indiscriminate and random nature of these violence are pertinent markers and they may derive from a plethora of triggers ranging from gang-related violence to political issues that arise post-elections.[37] The UN High Commissioner for Refugees' Guidelines on International Protection describes generalized violence as 'violence that is widespread, affecting large groups of persons or entire populations, serious and/or massive human rights violations, or events characterized by the loss of government control and its inability or unwillingness to protect its population - including situations characterized by repressive and coercive social controls by non-state actors, often pursued through intimidation, harassment and violence.' [38]

---

[36] *Cartagena Declaration on Refugees, Colloquium in the International Protection of Refugees in Central America, Mexico and Panama* (19–22 November 1984), UN Commission on Human Rights, Addendum, 'Guiding Principles on Internal Displacement' *Report of the Representative of the Secretary-General, Mr. Francis M. Deng, submitted pursuant to Commission on Human Rights resolution 1997/39,* UN Doc. E/CN.4/1998/53/Add.2 (11 February 1998).

[37] Cantor (2014, 34), Kinosian (2016), Simon (2016), Cantor and Plewa (2017).

[38] Guidelines on international protection no 12: claims for refugee status related to situations of armed conflict and violence under Article 1A (2) of the 1951 Convention and/or 1967 Protocol relating to the Status of Refugees and the regional refugee definitions (2016), para 58.

2.6 Development Projects

Development projects are a principal driver of internal displacement, however, not much research exists in this field. While it is often the case that other root causes of internal displacement emerge from situations that are in themselves a breakdown of societal order or which may be characterised as cataclysms such as armed conflicts, natural disasters and generalized violence, development projects are rather different. As implied in the development narrative, these projects often emerge from a legitimate aim to enable a beneficial agenda. Conceived as enablers of economic growth, development projects are initiatives that seek to advance public goods. However, intrinsic to this aim are two competing imperatives: the development project and the rights of those likely to be displaced. Across many societies, the imbalance often emerges in the primacy given to these projects over their impacts on those displaced. Consequently, issues of inadequate compensation, consultation, improper resettlement are often characteristic of how these projects are implemented. Research in the social sciences has flourished on this subject, shaping early knowledge formations and affording a useful optics and providing a conceptual scaffolding on which to diagnose the risks associated with a development paradigm that is not inclusive of project-affected populations. While these projects are in themselves not nascent phenomena in many African societies, there is yet to emerge a significant body of work in the context of the law on internal displacement. Specifically, within the African context, having studies that locates the discourse within national contexts are an imperative to spotlight the issue of development-induced displacement.

References

Ackerly B, Vandenbergh MP (2008) Climate change justice: the challenge for global governance. Georgetown Int Environ Law Rev 20:553–571

Adeola R (2015) Preventing xenophobia in Africa: what must the African Union do? Afr Hum Mobility Rev 1(3):253

Amber H (2011) Don't sell your coat: surprising truths about climate change. Lansing International Books, East Lansing

Barnett J, Campbell J (2010) Climate change and Small Island States: power, knowledge and the South Pacific. Earthscan, London

Bekker G (2009) Mass expulsion of foreign nationals: a "special violation of human rights" – Communication 292/2004 Institute for Human Rights and Development in Africa v Republic of Angola. Afr Human Rights Law J 9:262

Bell L (2011) Climate of corruption: politics and power behind the global warming hoax. Greenleaf Book Group Press, Austin

Berezin M (2006) Xenophobia and the new nationalisms. In: Delanty G, Kumar K (eds) The Sage Handbook of nations and nationalism. Sage Publications, Thousand Oaks

Booker C (2009) The real global warming disaster: is the obsession with 'climate change' turning out to be the most costly scientific blunder in history? Bloomsbury, London

Boykoff MT, Roberts TJ (2007) Media coverage of climate change: current trends, strengths, weaknesses (human development report office occasional paper, United Nations Development Programme

Burnett D (2014) Climate change is an obvious myth – how much more evidence do you need? The Guardian 25 November

Burns WCG (2000) The impact of climate change on pacific island developing countries in the 21st century. In: Gillespie A, Burns WCG (eds) Climate change in the South Pacific: impacts and responses in Australia, New Zealand and Small Island States. Kluwer Academic Publishers, Dordrecht

Campbell J, Bedford R (2014) Migration and climate change in Oceania. In: Piguet E, Laczko F (eds) People on the move in a changing climate: the regional impact of environmental change on migration. Springer, Cham

Cantor DJ (2014) The new wave: forced displacement caused by organized crime in Central America and Mexico. Refug Surv Q 33(3):34–68

Cantor DJ, Plewa M (2017) Forced displacement and violent crime: a humanitarian crisis in Central America? Humanitarian exchange no 69. Humanitarian Practice Network, London

Cartagena declaration on refugees, colloquium in the international protection of refugees in Central America, Mexico and Panama. 19–22 Nov 1984

Chestney N (2018) The world is at risk of reaching an irreversible "hothouse" state. World Economic Forum 9 August

Climate Change Secretariat (UNFCCC) (2005) Climate change: small island developing states. UNFCCC, Bonn. https://unfccc.int/resource/docs/publications/cc_sids.pdf

Crush J, Pendleton W (2004) Regionalizing xenophobia? Citizen attitudes to immigration and refugee policy in Southern Africa. Southern African Migration Project, Cape Town

Crutzen PJ, Steffen W (2003) How long have we been in the Anthropocene era? Climate Change 61(3):251–257

Daily Nation (2018) World Bank report says climate change to hit East Africa hard. Daily Nation 21 March

Darwall R (2019) Green tyranny: exposing the totalitarian roots of the climate industrial complex. Encounter Books, New York

Ddamulira R (2016) Climate change and energy in East Africa. Development 59(3–4):257–262

Deane P (1979) The first industrial revolution. Cambridge University Press, Cambridge

Decision 1/CP.16: The Cancun agreements: outcome of the work of the ad hoc working group on long-term cooperative action under the convention, adopted at the 16th session of the United Nations framework convention on climate change conference of the parties, Cancun, Mexico. 29 Nov–10 Dec 2010

Delanty G, Mota A (2017) Governing the Anthropocene: agency, governance, knowledge. Eur J Soc Theory 20(1):9–38

Egeru A, Nandozi C, Kyobutungi A, Tabuti JRS (2016) Dimensions of vulnerability to climate change shocks and variability in pastoral production systems of East Africa. In: Yanda PZ, Mung'ong'o CG (eds) Pastoralism and climate change in East Africa. Mkuki na Nyota Publishers Ltd, Dar es Salaam

El Hinnawi E (1985) Environmental Refugees. United Nations Environment Programme, Nairobi

Farmbry K (2019) Migration and xenophobia: a three country exploration. Lexington Books, Lanham

Ferris E (2014) Climate change is displacing people now: alarmists vs. skeptics. The Brookings Institution (Blog) 21 May

Foster GL, Royer DL, Lunt DJ (2017) Future climate forcing potentially without precedent in the last 420 million years. Nat Commun 8:14845

Gabbatiss J (2018) What is "Hothouse Earth", and how bad would such a climate catastrophe be? Independent (UK) 7 August

Global compact on safe, orderly and regular migration, adopted by the intergovernmental conference on the global compact, 10–11 Dec 2018

Gorwa R (2019) Poland: unpacking the ecosystem of social media manipulation. In: Woolley SC, Howard PN (eds) Computational propaganda: political parties, politicians, and political manipulation on social media. Oxford University Press, Oxford

Guidelines on international protection no 12: claims for refugee status related to situations of armed conflict and violence under Article 1A (2) of the 1951 Convention and/or 1967 Protocol relating to the Status of Refugees and the regional refugee definitions (2016)

Henckaerts JM (1995) Mass expulsion in modern international law and practice. Martinus Nijhoff Publishers, The Hague

Hoegh-Guldberg O (2012) Implications of climate change for Asian-Pacific coastal and oceanic environments. In: Warner R, Schofield C (eds) Climate change and the oceans: gauging the legal and policy currents in the Asia Pacific and beyond. Edward Elgar Publishing, Cheltenham

Hoffman AJ, Jennings DP, Lefsrud LM (2015) Climate change in the era of the Anthropocene – an institutional analysis. Michigan Ross School of Business Working Paper Working Paper 1280, June

Intergovernmental Panel on Climate Change (1990) Climate change: the IPCC impacts assessment. Australian Government Publishing Services, Canberra

Intergovernmental Panel on Climate Change (1992) Climate change: the IPCC 1990 and 1992 assessments. Intergovernmental Panel on Climate Change, Geneva

Intergovernmental Panel on Climate Change (2013) Working Group I contribution to the fifth assessment report of the Intergovernmental Panel on Climate Change. Cambridge University Press, Cambridge

Intergovernmental Panel on Climate Change (2014) Climate change 2014: mitigation of climate change. Contribution of Working Group III to the fifth assessment report of the Intergovernmental Panel on Climate Change. Cambridge University Press, Cambridge

Intergovernmental Panel on Climate Change (2018) Global warming of 1.5°C. Intergovernmental Panel on Climate Change, Geneva

Jacobson JL (1988) Environmental refugees: a yardstick of habitability. Worldwatch Institute, Washington, DC

Jakobeit C, Methmann C (2012) "Climate refugees" as dawning catastrophe? A critique of the dominant quest for numbers. In: Scheffran J, Brzoska M, Brauch HG, Link PM, Schilling J (eds) Climate change, human security and violent conflict: challenges for societal stability. Springer, Cham

James Bordeau J (2010) Xenophobia: the violence of fear and hate. The Rosen Publishing Group, Inc, New York

Johnson T (2010) Alternative views on climate change. Council on Foreign Relations 23 February

Kerridge E (1967) The agricultural revolution. Allen and Unwin, London

Kimuyu P (2016) The role of the IPCC in the understanding of climate change. GRIN Verlag, Munich

Kinosian S (2016) El Salvador's gang violence: turf wars, internal battles and life defined by invisible borders. Security Assistance Monitor (News), 10 February

Kirton JJ, Kokotsis E (2016) The global governance of climate change: G7, G20, and UN leadership. Routledge, Abingdon

Leroux M (2006) Global warming – myth or reality? The erring ways of climatology. Springer, Cham

Lewin-Epstein N, Levanon A (2005) National identity and xenophobia in an ethnically divided society. Int J Multicult Soc 7(2):90

Martin S (2010) Climate change, migration and governance. Glob Gov 16(3):397–414

Mathez EA (2009) Climate change: the science of global warming and our energy future. Columbia University Press, New York

Matt McGrath M (2018) Climate change: "'Hothouse Earth'" risks even if CO2 emissions slashed. BBC News 6 August

Mayer B (2011) Constructing "climate migration" as a global governance issue: essential flaws in the contemporary literature. McGill Int J Sustain Dev Law Policy 9(1):87–117

Mazoyer M, Roudart L (2006) A history of world agriculture: from the Neolithic age to the current crisis. Monthly Review Press, New York

Meyer S (2017) The first humans and early civilizations: the Neolithic Revolution. The Rosen Publishing Group Inc, New York

Michaels PJ, Balling RC (2009) Climate of extremes: global warming science they don't want you to know. Cato Institute, Washington, DC

Misago JP, Freemantle I, Landau L (2015) Protection from xenophobia: an evaluation of UNHCR's regional office for Southern Africa's xenophobia related programmes. United Nations High Commissioner for Refugees, Geneva

Mokyr J (n.d.) The second industrial revolution, 1870–1914. https://pdfs.semanticscholar.org/d3fc/63c43a656f01f021fb79526d9ba3b25f6150.pdf?_ga=2.47503399.1092564889.1585336886-649330839.1581418981

Munro A (2010) Climate change and SIDS. In: Nath S, Roberts JL, Madhoo YN (eds) Saving small island developing states: environmental and natural resource challenges. Commonwealth Secretariat, London

Myers N (1993) Environmental refugees in a globally warmed world. BioScence 43(11):752–761

Myers N (1997) Environmental refugees. Popul Environ 19(2):167–182

Myers N (2002) Environmental refugees: a growing phenomenon of the 21st century. Philos Trans R Soc 357:609–613

Neocosmos M (2010) From "foreign natives" to "native foreigners" explaining xenophobia in post-apartheid South Africa: citizenship and nationalism, identity and politics. Council for the Development of Social Science Research in Africa, Dakar

Nyamnjoh FB (2006) Insiders and outsiders: citizenship and xenophobia in contemporary southern Africa (Council for the Development of Social Science Research in Africa, Dakar

Overton M (1989) Agricultural revolution? England, 1540-1850. In: Digby A, Feinstein C (eds) New directions in economic and social history. Macmillan, London

Overton M (1996) Agriculture revolution in England: the transformation of the agrarian economy: 1500–1850. Cambridge University Press, Cambridge

Penprase BE (2018) The fourth industrial revolution and higher education. In: Gleason NW (ed) Higher education in the era of the fourth industrial revolution. Palgrave Macmillan, London

Piguet E (2013) From "primitive migration" to "climate refugees": the curious fate of the natural environment in migration studies. Ann Assoc Am Geogr 103:148–162

Poole S (2018) From greenhouse to hothouse: the language of climate change. The Guardian (UK) 9 Aug

Pouch A, Zaborska A (2015) Climate change influence on migration of contaminants in the artic marine environment. In: Zielinski T, Weslawski M, Kuliński K (eds) Impact of climate changes on marine environments. Springer, Cham

Ribot J (2014) Cause and response: vulnerability and climate in the Anthropocene. J Peasant Stud 41(5):667–705

Rifkin J (2012) The third industrial revolution: how the internet, green electricity, and 3-D printing are ushering in a sustainable era of distributed capitalism. The World Financial Review 3 March

Roemer JE, van der Straeten K (2006) The political economy of xenophobia and distribution: the case of Denmark. Scand J Econ 108(2):251–277

Rydgren J (2004) The logic of xenophobia. Ration Soc 16(2):123–148

Saideman SM, Ayres RW (2008) For kin or country: xenophobia, nationalism and war. Columbia University Press, Columbia

Scharre P (2018) Army of none: autonomous weapons and the future of war. WW Norton and Company, New York

Schurer AP, Mann ME, Hawkins E, Hegerl GC, Tett SFB (2017) Importance of the pre-industrial baseline in determining the likelihood of exceeding the Paris limits. Nat Clim Chang 7(8):563–567

Schwab K (2017) The fourth industrial revolution. Crown Business, New York

Schwab K, Davis N (2018) Shaping the future of the fourth industrial revolution: a guide to building a better world. Penguin, London

Simon TJ (2016) Gang-based violence and internal displacement in El Salvador: identifying trends in state response, human rights violations, and possibilities for asylum: policy analysis exercise. http://www.nrc.org.co/wp-content/uploads/2017/11/gang-based_violence_and_internal_displacement_in_el_salvador-_identifying_trends_in_state_response_human_rights_violations_and_possibilities_for_asylum.pdf

Skilton M, Hovsepian F (2018) The 4th industrial revolution: responding to the impact of artificial intelligence on business. Palgrave Macmillan, London

Smith R, Free M (2016) The great disruption: competing and surviving in the second wave of the industrial revolution. St Martin's Press, New York

Snowden E (2019) Permanent record. Macmillan, London

Stager CJ (2016) Climate change in the age of humans. In: Sample AV, Bixler PR, Miller C (eds) Forest conservation in the Anthropocene: science, policy and practice. University Press of Colorado, Colorado

Suhrke A (1993) Pressure points: environmental degradation, migration and conflict. American Academy of Art and Science, Boston

Thornton F (2018) Climate change and people on the move: international law and justice. Oxford University Press, Oxford

UN Commission on Human Rights, Addendum. Guiding principles on internal displacement. Report of the Representative of the Secretary-General, Mr. Francis M. Deng, submitted pursuant to Commission on Human Rights resolution 1997/39, UN Doc. E/CN.4/1998/53/Add.2 (11 Feb 1998)

UN Committee on the Elimination of Discrimination against Women. Joint general recommendation no. 31 of the Committee on the Elimination of Discrimination against Women/general comment no. 18 of the Committee on the Rights of the Child on harmful practice, UN Doc CEDAW/C/GC/31-CRC/C/GC/18 (14 Nov 2014)

UN Human Rights Council, Report of the Office of the United Nations High Commissioner for human rights on the relationship between climate change and human rights, annual report of the United Nations high commissioner for human rights and reports of the Office of the High Commissioner and the Secretary-General, UN Doc A/HRC/10/61 (15 Jan 2009)

United States Global Change Research Program (2017) Climate science special report: fourth national climate assessment, volume I. US Global Change Research Program, Washington, DC

Watts, J (2018) Domino-effect of climate events could move Earth into a "hothouse state". The Guardian (UK) 7 August

Wright C, Nyberg D, Rickards L, Freund J (2018) Organizing in the Anthropocene. Organization 25(4):455–471

Yakushko O (2009) Xenophobia: understanding the roots sand consequences of negative attitudes towards immigrants. Couns Psychol 37(1):36–66

Zalasiewicz J, Williams M, Hearing T (2018) Hothouse Earth: our planet has been here before – here's what it looked like. The Conversation 13 August

Chapter 3
Emerging Issues and Internal Displacement in Africa: Interlinkages

Abstract This chapter examines the nexus between the emerging issues and internal displacement, exploring the intersections and evidence that reflect interlinkages. The aim of this chapter is to respond to the knowledge gap required for the furtherance of discussion on these issues in the context of internal displacement.

Keywords Internal displacement · Internally displaced persons · Climate change · Technology · Xenophobia · Harmful practices · Generalized violence · Development projects

3.1 Climate Change

The global rhetoric on climate change reflects an imminent challenge that needs to be resolved. Under the *Paris Agreement* 2015, countries agreed to limit global temperature to 1.5 °C while sustaining 'the increase in the global average temperature to well below 2 °C above pre-industrial levels'.[1] However, reaching this target has become rather bleak given the slow progressions towards the target and uneven political will among states on climate change. In 2019, the United Nations Environment Programme (UNEP) reported that[2] had serious climate action begun in 2010, the emissions reductions required per year to meet the emissions levels in 2030 consistent with the 2 °C and 1.5 °C scenarios would only have been 0.7 per cent and 3.3 per cent per year on average. However, since this did not happen, the required cuts in emissions are now 2.7 per cent per year from 2020 to year- 2030 for the 2 °C goal and 7.6 per cent per year on average for the 1.5 °C goal.

Consequently, the report emphasises that 'quick wins' are needed or 'the 1.5 °C goal of the Paris Agreement will slip out of reach.'[3] What this point reinforces is the fact that climate change impacts are increasingly imminent. Moreover, with the increase in global temperatures, one of such ramifications of impact for which sus-

---

[1] Paris Agreement (2015), art 2(1)(a).

[2] United Nation Environment Programme (2019, 26).

[3] United Nation Environment Programme (2019), Hausfather (2019).

tainable solutions are required is internal displacement. In 2018, the World Bank estimated that more than 140 million people would move internally within countries worldwide. Of this figure, at least 86 million people will move internally within sub-Saharan African states due to climate change.[4] Although the data conflates mobility more broadly with internal displacement, the projection reinforces the need for adequate attention to climate change as an IDP research agenda.

Four dimensions of the nexus between climate change and internal displacement can be discerned.[5] The first dimension is sudden on-set disasters. Sudden on-set disasters are often cataclysmic, resulting in violent impacts and causing large movements of populations.[6] In the narrative, this form of disaster is often portrayed through massive population displacement due to torrential rainfalls.[7] In determining the impact of climate change on rainfall patterns, scientists often look to recurrence intervals as a measure of ascertaining change in frequency by evaluating the return cycle of past rainfall episodes given that the information offers valuable understanding of the impact of climate variations. To ascertain if a rainfall episode is extreme, climatologists consider whether the intensity of the rainfall over a period of time returns in a period greater or equivalent to 5 years for any sample episode between 5 minutes and 24 hours.[8] Although studies have revealed that warming conditions in West Africa have increased between 0.3 and 1 °C, these patterns have been more connected with dry extremes.[9] As observed by Zahiri et al. and others, '[t]he underlying causes of such inundations are diverse, complex, and involve not only the absence of efficient rainwater drainage systems, uncontrolled urban expansion, and the construction in water outlets but also the occurrence of extreme rainfall events or extreme rainfall intensities.'[10]

However, a significant body of research has emerged on the impact of climate change on slow on-set disasters in Africa. Unlike sudden onset disasters, this form of disaster builds over a period of time.[11] They include gradual events such as: sea level rise, desertification, forest degradation, loss of biodiversity and ocean acidification. A notable paradigm that exemplifies this scenario is the movement of pastoralists in West and East Africa. These populations are increasingly being recognised

---

[4] World Bank (2018b).

[5] See Kälin (2010, 84–91), McAdam (2012, 18–19).

[6] Hugo (2011, 225, 261).

[7] In the southern African region, the intensity of cyclones and torrential rainfall patterns have been linked to climate change. Yarnell and Cone observe that 'scientists representing an overwhelming consensus in the scientific community have concluded that cyclone intensity and rainfall rates are expected to increase. Southern Africa must be prepared for more extreme and unusual weather.' Evidently, the displacement situation brought about by Cyclone Idai in parts of Zimbabwe and Mozambique lends credence to the pertinence of underscoring this linkage. See Yarnell and Cone (2019).

[8] See Casas et al. (2004, 139–150), Zahiri et al. (2016, 15, 23).

[9] Sylla et al. (2016, 25–40), Zahiri et al. (2016, 16).

[10] Zahiri et al. (2016, 15–16).

[11] Bryne (2018, 765–766).

as "climate canaries" fated to be dismally affected by climate change.[12] Their cyclical and seasonal movements are increasingly becoming disrupted due to slow onset events such as rising temperatures in East Africa. Leveraging climate models, scientists have drawn links between droughts conditions in the region and climate change which have induced pastoralist movements.[13] Climate-induced stresses on water points and grazing lands are also affecting movement patterns of these populations. Given the gradual depletion of these resources, pastoral populations, traditional hunters and gatherers are becoming increasingly at risk of poverty and vulnerable to prolonged displacements.[14]

The third dimension is climate-related conflicts. This dimension relates to the scarcity of resources due to climate-induced stresses that in turn result in conflict between communities that share these resources.[15] Inter-ethnic clashes in the northern parts of Nigeria between transhumance populations and agrarian communities illustrates this scenario.[16] Frequent droughts and desertification have notably induced north-south movement of herdsmen within the country in search of grazing territories,[17] consequently brewing conflicts. Due to these conflicts, significant population displacements have occurred.[18] Similar narratives have also emerged in parts of Ghana, Ivory Coast and Burkina Faso.[19]

The fourth dimension relates to climate development-induced displacement. This dimension is mostly a consequence of responses to climate change, i.e. adaptation and mitigation.[20] While there are various scenarios of this form of displacement globally, three pertinent instances in Africa are discussed. The first is the case of the Sengwer peoples of Kenya. With support from the World Bank for REDD+ readiness,[21] the Kenyan government developed the Natural Resource Management Programme (NRMP) aimed at building capacity towards afforestation, forest conservation, managing water catchments and preventing degradation.[22] Based on the fact that the project would affect indigenous peoples, an Indigenous Peoples'

---

[12] Jegede (2017, 172).

[13] In a scientific research on 2011 East African drought and climate attribution, 'evidence was found for an enhanced risk of failure of the long rains in 2011 due to human-induced climate change'. See Lott et al. (2013, 1177–1181), Straziuso (2013), Sengupta (2018).

[14] Adaptive migration has become a coping mechanism among forest populations in parts of Central and East Africa where 'lengthy dry season, are affecting the agricultural calendar and bringing about a scarcity of forest products, such as fruits and tubers of plants such as potatoes and yam'. See Jegede (2017, 172).

[15] Reuveny (2007, 656–673), Freeman (2017, 351–374).

[16] See International Crisis Group (2017, 3), International Crisis Group (2018).

[17] Fasona and Omojola (2005); Amusan et al. (2017, 35).

[18] Aluko (2018), Alindogan (2018), Amnesty International (2018b).

[19] Cabot (2015, 159–160), Olaniyan et al. (2015, 53).

[20] Gausset and Whyte (2012, 214), Mabikke (2014, 142).

[21] World Bank Inspection Panel (2013, para 100).

[22] World Bank (2007, 1).

Planning Framework (IPPF) was prepared to promote an inclusive decision-making process and ensure that indigenous populations are protected.[23] However, the objective of the IPPF was not fully complied with. Homes belonging to the Sengwer indigenous peoples in the Embobut forest were torched by the Kenya Forest Services leading to the displacement of thousands of Sengwer peoples.[24] In Uganda, the case of the New Forest Company (NFC) is also illustrative of this point. More than 20,000 people were dispossessed of their land without adequate compensation for a timber plantation by the NFC in Uganda. The project, which was meant to generate carbon credits while also exporting timber in commercial quantity, had the consequence of destroying properties of local farmers without adequate compensation.[25] Another pertinent example of this scenario is the case of the Ibi-Batéké Carbon Sink Plantation Project in the Democratic Republic of Congo (DRC). The project was developed to promote reforestation of 4220 hectares on the Batéké Plateau, produce sustainable supply of fuelwood to Kinshasa and increase carbon sinks that could sequestrate over two million tons of $CO2$ over a 30-year period.[26] Although an impact assessment was carried out prior to the implementation project with 'an extensive consultation'[27] with communities and local authorities, the Batwa indigenous populations in the area were not properly involved in the consultation processes. Moreover, their lands were plundered inducing displacement.[28] As with the case of the Batwa, issues of inadequate consultation have also emerged in the treatment of local populations in other parts of the continent. Some communities have also begun resisting initiatives geared towards climate change mitigation and adaptation in view of the impact of displacement.[29]

3.2 Technology

The positive impacts of technology are without doubt replete. The fact that it makes connectivity faster, advances globalization, drives economies, and enhances service delivery are some of the pertinent imperatives for which it has become globally

---

[23] Schmitd-Soltau (2006).

[24] Langat (2014), Amnesty International (2018a, 7–8).

[25] Grainger and Geary (2011).

[26] World Bank (2009), Mushieta and Merrill (2010, 352).

[27] Adrien (2007, 65).

[28] Adrien (2007, 61).

[29] With regards to REDD readiness project in the Rufiji Delta of Tanzania, Beymer-Ferris and Bassett observe that 'the Warufiji [villagers] are resisting efforts to make the Rufiji Delta North "REDD ready" on the grounds that these efforts will increase their vulnerability and displacement.' Beymer-Ferris and Bassett (2012, 339).

relevant.[30] However, the fact that technologies are also simulating human activities and making human decisions has also brewed a context for concern. In the context migration and mobility, concerns have emerged on the deployment of technology in migration governance, particularly in border control and management. In the context of internal displacement, the nexus between technology and IDP protection is yet to emerge in research and in law and policy on internal displacement. However, emerging evidence suggests that this nexus is not entirely futuristic in its impact given the significant advancement in technology, in particular artificial intelligence and machine learning.

There are two identifiable intersections between technology and internal displacement for which pertinent solutions are imperative globally, and in the African context. The first relates to technology-enabled attacks. A pertinent instance of this form of internal displacement, for instance, is in the Somalia context where airstrikes have been forcing civilian populations to flee their homes. In 2019, *Foreign Policy* reported that 'some 450,000 people have been displaced from al-Shabab strongholds in the Lower and Middle Shabelle regions that frame Mogadishu, the coastal capital, where the United States is responsible for air operations, according to nongovernmental organization and United Nations agencies that operate in the area'.[31] These airstrikes, a consequence of both manned and unmanned aircrafts have increasingly become regular fixtures in countering insurgencies and are increasingly shaping the future of military combats in many parts of the world. While manned aircrafts are generally within the control of a human pilot with the capability of making decisions, the rhetoric of military operations with the use of unmanned aerial vehicles (UAVs), where such vehicles are completely autonomous creates a significant concern with regards to the safety of civilian populations. Although these UAVs colloquially known as drones, have significant benefits in military operations such as 'potential low-observability, significant flight endurance, reuse on future missions if successfully launched and recovered, and even

---

[30] In a 2019 report, the United Nations Commission on Science and Technology for Development reinforced the pertinence of technology in driving the realization of the sustainable development goals. According to UNCTAD, 'Frontier technologies, including big data and machine learning, can also be used to create, measure and develop and monitor more broadly the effectiveness of development programmes and progress towards the Sustainable Development Goals. Models based on both mobile telephone activity and airtime credit purchases have been shown to estimate multidimensional poverty indicators accurately, while recent studies have validated the potential of satellite imagery and machine learning in estimating household consumption and assets, using publicly available and non-proprietary data'. United Nations Commission on Science and Technology for Development (2019, para 7).

[31] Sperber (2019).

plausible deniability of ownership, depending on the system',[32] these benefits equally unveil some of the downsides of these technologies.[33]

Another dimension of technology and internal displacement is technology-enabled generalized violence. This dimension emanates from the growing impact of technology, particularly social media, in fuelling violence. With the widespread prevalence of fake news, deep fakes and false content on social media platforms, the endless potential of technology as an enabler of violence and consequently, displacement is evident. This dimension may reflect, for instance, in generalized violence enabled by xenophobic or extremist rhetoric spread through online platforms. Prior to a wave of xenophobic attacks in South Africa, it was observed that anti-migrant sentiments had been spreading through social media. According to Burke 'community leaders' had indicated that the attacks had 'been building for several months. Anti-immigrant rhetoric has been circulating on social media and among groups who allege immigrants cheat their customers with out-of-date produce in their shops, take jobs from locals and defraud the state.'[34] The pertinence of paying attention to this dimension emanates from the fact that it is not often visibly connected in law and policy formations given the dearth of studies in this regard. But given the growing impact of technology and the emerging nexus between technology and human rights being more visibly drawn, giving attention to various dimensions of its impacts is pertinent.

3.3 Xenophobia

There are two pertinent dimensions of the nexus between xenophobia and internal displacement. The first resonates from its nexus with generalized violence as a precipitant of such violence: xenophobia-induced generalized violence. The second

---

[32] Brookes (2019, 2), https://www.heritage.org/sites/default/files/2019-09/BG3437.pdf (accessed 28 March 2020).

[33] While information on the impact of drone attacks in the African contexts, have been scarce, in part, due to the unavailability of accurate data and evidence, some of these impacts are well-documented in other regions. For instance, the UN Special Rapporteur on the Promotion and Protection of Human Rights and Fundamental Freedoms while Countering Terrorism has spotlighted the negative impact of these technologies on civilian populations in the Middle East region. With respect to Pakistan, for instance, the Special Rapporteur observed that '[o]n 30 October 2006, precision-guided munitions were reportedly fired at a religious seminary in Chenagai in the Bajaur tribal region. Remotely piloted aircraft under the control of the United States are alleged to have been involved in the operation. Up to 80 people were reportedly killed instantly during the attack; two more victims reportedly died in hospital shortly afterwards as the result of injuries sustained. It is alleged that as many as 69 of the dead were children under 18 years of age, and that 16 of those killed were under the age of 13. Eyewitnesses allege that the majority of those killed had been pupils at the seminary and were non-combatant civilians.' See UN Human Rights Council (2014, para 47).

[34] Burke (2019b).

relates to the implementation of discriminatory policies against specific groups of foreign nationals.

Over the last decade, the nexus between xenophobia and generalized violence has become an issue of concern in many societies. In Africa, this concern has emerged in some societies with increasing calls for action. With respect to South Africa, for instance, this issue has gained significant attention. Although pockets of attacks emerged in the late 1994 towards 1995 in Alexandra, only in the 2000s did these become more frequent. In this period, attitudinal prejudices towards foreign nationals began to emerge on the rhetoric of jobs and criminality. Mostly in this period, Zimbabweans were attacked with some displaced.[35] In May 2008, the first xenophobic attack occurred. This was started by riots against non-nationals in the township of Alexandra.[36] The image of a Mozambican burnt to death in this episode of attack significantly drove global shock. However, in 2 weeks, over 60 people were killed and at least 100,000 people were displaced.[37] The rhetoric of jobs and crimes significantly fuelled further attacks in subsequent years resulting in the displacement of significant populations in various parts of the state. In 2009, at least 3000 Zimbabweans were displaced in Western Cape.[38] Aside from incidences of widespread lootings, horrific killings have also emerged in this context. In 2015, a significant wave of attacks emerged following comments by the Zulu King that foreigners should return home 'because they are changing the nature of South African society with their amanikiniki or goods and enjoying wealth that should have been for local people'.[39] In April 2015, at least 5603 people who were displaced in KwaZulu-Natal were being 'temporarily housed in shelters outside the areas affected by the violence'.[40] While a majority were Malawians, Zimbabweans and Mozambicans were also affected.[41] While much of these attacks have been fuelled by the attitudinal prejudices of local populations, the rhetoric from the state reinforcing these prejudices, the failure of the government to address the failure of the criminal justice system and the systemic avoidance of the root causes of economic unrest has significantly propelled the continuance of the vice. In 2019, following xenophobic attacks, the UN Refugee Agency reported that not less than '1500 foreign nationals, predominantly migrants but also refugees and asylum-seekers, have been forced to flee their homes.'[42]

Another dimension of the nexus between xenophobia and internal displacement relates to situations where displacement occurs due to the implementation of discriminatory policies against specific groups of foreign nationals. An example of

---

[35] Crush et al. (2017b, 21).

[36] Crush et al. (2017a, 5).

[37] Chaskalson (2017).

[38] The News Humanitarian (2009), UNHCR (2009), South African History Online (2015).

[39] Hans (2015).

[40] Provincial Government of KwaZulu-Natal (2015, 119).

[41] Crush et al. (2017b, 25).

[42] UNHCR (2019).

this, for instance, was in Kenya.[43] In December 2012, the government of Kenya issued a press release in which it expressed the fact that hosting large refugee populations 'has not been without challenges. Among them, a major challenge has been rampant insecurity in the refugee camps and urban areas.'[44] It further expressed the fact that 'in this public domain many people have been killed and several more injured with grenade attacks in our streets, churches, buses and in business places.'[45] Consequently, in view of 'this unbearable and uncontrollable threat to national security, the government has decided to put in place a structure encampment policy'.[46] For the implementation of this, the government emphasised that 'asylum seekers and refugees from Somalia should move back to Dadaab refugee camps; ... [and] asylum seekers and refugees in the urban areas from other countries should move back to Kakuma refugee camp'.[47] Moreover, the government further decided that 'registration of asylum seekers and refugees in urban areas ... [be] stopped and all registration centers closed'.[48] Although these refugees possessed documentation to reside in urban areas, the Acting Commissioner for Refugee Affairs expressed that consequent on the order 'their legal documentation ... [had] ceased to function in the urban areas. So, if they continue staying in the urban areas, then they will be staying illegally and that is a function of another department of government, probably police and immigration.'[49] Consequently, security operations were set in motion to enforce this policy which resulted in the forced relocation of thousands of Somalis into refugee camps in Dadaab and Kakuma.[50] The rhetoric of increased insecurity due to large populations of refugees and the consequent imperative for population movement has also emerged in other national contexts such as Tanzania.[51] In 2001, for instance, the former President of Tanzania, Benjamin Mkapa urged refugees to go back into camps.[52] This was set against the clime of insecurity.[53] In another press conference, the former President had expressed to a mission of the UN Security Council that he was tired of 'perennial accusations that his country was

---

[43] Opiyo and Githinji (2011), Human Rights Watch (2012b); Abdi (2012), Wambua-Soi (2012), Nzwili (2012), BBC (2012), Odula (2012), Relocation of urban refugees to officially designated camps, Letter from E Mutea Iringo, Permanent Secretary, Provincial Administration and Internal Security (2013), Human Rights Watch (2013), Reini (2013), Human Rights First (2013), Rinelli and Opondo (2015, 138), Lind et al. (2017, 118), Human Rights Watch (2012a).

[44] Department of Refugee Affairs (2012).

[45] As above.

[46] As above, 1.

[47] As above, 2.

[48] As above, 2.

[49] Kenya Citizen TV (2012).

[50] Yarnell 2014.

[51] Xinhua News Agency 2001.

[52] As above.

[53] As above.

harbouring rebels from Burundi'[54] and that: 'once these refugees leave we will be able to live in peace and continue with our development activities'.[55]

3.4   Harmful Practices

In understanding the nexus between harmful practices and internal displacement, it is relevant to examine these practices and populations that have been displaced as a consequence. The first set of practices are: gender-based practices. These practices relate to those against women and girls. Four main practices are discussed in this context: sexual servitude, breast ironing, female genital mutilation (FGM) and child marriage.

Sexual servitude (*trokosi*) is a practice used to serve rural justice in some parts of Ghana.[56] In this context, young girls are offered to shrines as compensation for crimes committed by male family relative. According to estimates 'between 4000 and 6000 women and children [are] under bondages in shrines in Ghana alone.'[57] Evidence of this practice has also emerged in other neighbouring countries such as Togo and Benin. There have been incidences of young girls fleeing such practices.[58] However, fleeing these situations have been associated with dire consequences. Gillard observes that '[g]irls who do try and escape and return to the village of their families are forcefully returned, as there is a penalty of fine (imposed on them by the fetish priests or elders of the villages) if they help the escapee.'[59] Moreover, Gillard notes that 'the very real fear that the curse will remain on the family, forces family members to return the child to the shrine, where they are beaten for trying to flee. For them, life is hopeless.'[60]

Breast ironing is another practice perpetrated against girls in the early stages of puberty. While it exists in various parts of West and Central Africa, much of the evidence on this practice has surfaced in Cameroon where estimates are 'as high as one in three [girls] (around 1.3 million)'.[61] This practice is premised on the desire to delay the physical development of girls to deter attention. Carried out with pestles, stones, wooden spoons, spatulas, coconut shells, or hammers that are heated and used to press the breasts of young girls,[62] breast ironing often has serious health

---

[54] The New Humanitarian (2001).

[55] As above.

[56] Bilyeu (1999, 457), Mistiaen (2013), Klein (2014, 370), Boaten (2001, 91), Akonor (2019).

[57] Mistiaen (2013).

[58] Farah (2001), *Lorraine Fiadjoe v. Attorney General of the United States* (2005), Kerry Kennedy Cuomo, Camera Works. https://www.washingtonpost.com/wp-srv/photo/onassignment/truth/st/09.htm.

[59] Gillard (2010, 14).

[60] Gillard (2010, 14).

[61] Bradley (2019).

[62] Pemunta (2016, 335).

implications on girls. This form of violence is perpetuated in private spaces by close female relatives, particularly mothers,[63] in hopes to 'make their daughters less sexually attractive to men, staving off early marriage and pregnancy, and keeping them in school'.[64] Hence, the rhetoric of good: that it is carried out in the girls' interest.[65] There have also been instances of girls fleeing their homes to escape these practices,[66] sometimes exposed to other forms of abuses.[67]

Linked to perceptions of femininity and marriageability, FGM affects millions of girls across the continent and has been recorded in communities across 29 African countries.[68] FGM/cutting are surgical removals/cutting of all or part of the female genital organ. This practice has gained significant attention in laws and policies across the world as violence against women and girls and has also gained recognised as a ground for refugee status. Within national territories, there have also been instances of girls fleeing this practice.[69]

Child marriage is another prevalent harmful practice in many African societies. In some parts of the continent, it has resulted in the displacement of young girls.[70] With some of the highest prevalence in the world, it is estimated that Africa loses billions of dollars to child marriage economically.[71] The prevalence of this practice is mostly borne out of culture, poverty and religion. In Niger, where the prevalence of child marriage is at 76%,[72] poverty and the belief that marrying off girls early will prevent promiscuity, accounts for the prevalence of this practice.[73]

The second set of practices are group-based practices. These are practices perpetuated against specific groups based on unique features such as persons living with albinism. In parts of the continent, persons living with albinism have been forced to flee their homes given attacks on them for ritual purposes.[74] While this practice has emerged in countries such as Malawi and Burundi, [75] there have been significant evidence with regards to Tanzania. In 2009, thousands of persons with albinism were forced to flee their homes.[76] There are other group-based harmful practices with potential for causing internal displacement such as those perpetrated

---

[63] Koigi (2017), Selby and Ngalle (2018), Obaji (2020).

[64] Selby and Ngalle (2018).

[65] Selby and Ngalle (2018).

[66] Tetchiada (2006), Fem (2011).

[67] Tetchiada (2006).

[68] Equality Now (2019).

[69] Sanghani (2015), BBC (2016), Kiptoo (2016), FIGO (2018), UN Women (2019), BBC (2019), Jidovanu and Otieno (2019).

[70] Kyle Almond (2017); Okiror (2018).

[71] Wodon et al. (2018), World Bank (2018a).

[72] Ennaji (2019).

[73] UNICEF (2018).

[74] UN Human Rights Council (2017, para 52), Seepersaud (2017, 116).

[75] Dixon (2017).

[76] CNN (2009).

against specific groups in view of some physical features. A notable example for instance is the ritual killing of bald men in Mozambique. In 2017, reports surfaced indicating that bald men were being targeted for ritual killings in the state and about five of these men had been killed. These killings were premised on the superstition that the heads of bald men had 'gold'.[77] And as such, they were being hunted for ritual purposes. While these are reported cases, others often go unreported in many parts of the continent. However, drawing this nexus is imperative in fostering durable solutions to internal displacement.

3.5 Generalised Violence

There are five pertinent dimensions of the interlinkage between generalized violence and internal displacement in Africa: gang violence, electoral violence, inter-ethnic clashes, violent extremism and technology-enabled generalized violence. While much of the discussion on gang violence has centred on the activities of groups in the Latin America region, instances of these forms of violence has emerged in parts of the continent including South Africa, Kenya and Nigeria. In South Africa, for instance, the activities of gangs in the Western Cape province is a case in point.[78] In Nigeria, this form of violence has emerged with respect to the activities of criminal gangs in the south-western state of Lagos and in the north-western state of Sokoto.[79] In Kenya, gang violence has led to the displacement of residents, for instance, in the Mathare slum.[80]

In the context of electoral violence, the nexus between generalized violence and internal displacement has been pronounced. Notable examples of these resonate from post-election violence in countries such as Côte d'Ivoire and Kenya. In Cote d'Ivoire, the post-election violence emerged from a power tussle between the Laurent Gbagbo and Alassane Ouattara. Following the defeat of Gbagbo, who at the time was the incumbent president, election results were nullified in parts of the northern region of the country. The Constitutional Council of the country declared Gbagbo the winner of the elections.[81] However, conflict broke out shortly between loyalists of both parties which led to the death of around 3000 people and the displacement of at least 500,000 people in 2010.[82] In Kenya, the 2007/2008 post-election violence is another case in point. Following the presidential election in late 2007, incumbent Mwai Kibaki was declared the winner in an election that showed early signs of victory for the opponent, Raila Odinga. This subsequently resulted in

---

[77] BBC (2017).

[78] Morris (2019), Burke (2019a); Bax et al. (2019); de Greef (2019).

[79] The Guardian (2018), Al Jazeera (2019).

[80] Gettleman (2006).

[81] Rice (2010), *BBC* (2010), Zounmenou and Lamin (2011, 6).

[82] UN Office for the Coordination of Humanitarian Affairs (2011).

violence in parts of the country, including Nyanza, Rift Valley and Central Provinces. This subsequently resulted in the death of over 1000 people and the displacement of hundreds of thousands, placed between 300,000 and 650,000.[83]

Aside from electoral violence, interethnic clashes in various parts of the continent have also reinforced the nexus between generalized violence and internal displacement. A significant case in point is the inter-ethnic violence in Ethiopia that resulted in the displacement of millions of people.[84] Solely in 2018, nearly 2.9 million people were newly displaced due to inter-ethnic conflicts.[85] At least 800,000 people, mostly ethnic Gedeos were displaced from West Guji in Oromia due to these inter-ethnic conflicts, principally over land.[86] Similar patterns of conflict have also emerged in other parts of the continent including South Sudan and Mali.

Besides inter-ethnic clashes, violent extremism has significantly reinforced the link between generalized violence and internal displacement. These extremisms are replete in many parts of the continent. Among the famous forms is the Boko Haram insurgency in Nigeria and Al-Shabaab Somalia.[87] The Boko Haram conflict which began in 2009 has led to the internal displacement of more than two million people, mostly in the north-eastern part of Nigeria.[88] Rooted in a mix of ethnic/clan sentiments and religion, these violent extremisms have become a significant face of conflict in Africa with millions of people internally displaced by the activities of these groups. However, there are other new forms of violent extremism that have found expression in global extremist movements such as the new Al-Shabaab in Mozambique which has displaced about 100,000 people.[89]

Another significant nexus is the issue of technology-enabled generalized violence.[90] It is pertinent to emphasise that technology may also reinforce other forms of internal displacement. With the global increase in the use of technology and access in various societies, it is imperative to significantly advance protection for persons within this context. Moreover, to advance this imperative, there is a need to build knowledge on this dimension.

---

[83] Human Rights Watch (2017), UN High Commissioner for Human Rights (2008, 12), Human Rights Watch (2008).

[84] Gardner (2019b), Internal Displacement Monitoring Centre (2018).

[85] Internal Displacement Monitoring Centre (2019, 14–15), Gardner (2019a).

[86] Gardner (2019a), Obulutsa (2018).

[87] Kleinfeld (2020).

[88] UNHCR (n.d.).

[89] Kleinfeld (2020).

[90] Parts of this issue has been discussed in this chapter under the section on technology.

3.6 Development Projects

Development projects as significant drivers of internal displacement have emerged as a concern globally.[91] With the global estimate that at least 15 million people are displaced annually due to development projects, there is a need to enhance sustainable development and give attention to the plight of project-affected persons. It is within this context that a focus on this root cause is imperative. There are five pertinent dimensions of the nexus between development projects and internal displacement. These are: dams, natural resource extraction, agricultural investment, urban renewal and climate-related projects.

Dams have been the primary focus of research on the nexus between development projects and internal displacement.[92] This is due, in part, to the fact that this form of displacement has been most visible and dates back as early as the 1950s and 1960s with the Aswan High Dam in Egypt and the Akosombo dam in Ghana respectively. Given the value of hydropower energy in furthering economic activities within states, these projects are advanced as drivers and enablers of these activities.

These issues have also emerged in the context of natural resource extraction.[93] Extraction of resources such as oil, gas and gems have significantly led to the displacement of populations in many parts of the continent. While a significant body of research have examined oil extraction in Nigeria's Niger Delta, there are other examples of these from extraction of resources in other contexts, for instance, gold from Guinea, diamond from Zimbabwe and coal from Mozambique. With the continuing discovery of resources across many African countries and the exploration of existing resource fields, the continuing presence of this form of internal displacement is evident.

Moreover, agricultural investment projects have also emerged as an interlinkage between development projects and internal displacement.[94] Given the drive towards transforming economies and realising economic growth, states have adopted modernised forms of agriculture involving large-scale mechanised investments from subsistence to commercially viable modes to enhance productivity. The increasing acquisition of land by investors in many African countries to produce food have also raised displacement concerns. While such acquisitions for food production are driven by an imperative to address food insecurity and shortages in other regions, the manner in which such acquisitions are done and people are displaced to make way for these projects (often associated with the rhetoric of 'land grab'), raises tensions.[95]

---

[91] Adeola (2021), See also Cernea (1996, 13), Oliver-Smith (2005, 189), de Wet (2006, 180), Penz et al. (2011), Downing and Garcia-Downing (2009, 225).

[92] Tamakloe (1994, 99), Fahim (1981, 59), Scudder (2005), DeGeorges and Reilly (2006, 633), Thamae (2006).

[93] Madebwe et al. (2011, 292), Oluyemi (2014, 28), West African Democracy Radio (2017).

[94] FIAN (2012); FIAN International (2012).

[95] Kachika (2010, 20), Cotula (2013), Makochekanwa (2014, 58), Batterbury and Ndi (2018, 573).

The nexus between development projects and internal displacement also resonates within the context of urban renewal.[96] These projects are often encased in the rhetoric of clearing of slums and cleaning of cities. These projects have emerged as visible forms of displacement in Africa with thousands forced to move, mostly urban poor. Moreover, the increasing importance of addressing climate change has also begun to occasion situations of population displacement.[97] The need for climate adaptation and mitigation and the consequent impact of these strategies on populations made to make way for these projects unveils the rhetoric inherent in this form of internal displacement.

3.7 Conclusion

This chapter examines the nexus between six emerging issues and internal displacement in Africa, namely: climate change, technology, xenophobia, harmful practices, generalized violence and development projects. The link between climate change and mobility has become an emergent imperative in discussions on the global governance of climate change. There are studies that suggest that climate change and population movements are evident concerns. However, the normative response has been rather mixed and mostly located in the context of soft norms. While much of this is premised on the nature of the global migration discourse, the mixed rhetoric on the impact of climate change has also led to a slow-paced response. However, the imminence of the climate emergency, not least, the growing concerns of global average temperatures and states lagging behind the Paris Agreement has significantly reinforced the need for sustainable solutions to the various dimensions of climate change impact. Moreover, within the African regional context, normative progressions on climate change and internal displacement have emerged, reinforcing the need for attention to this issue. While there is a need for continuous knowledge formations on the nexus, notably, from a scientific viewpoint, there is a growing body of evidence-based knowledge on this relationship precipitating the need for legal response. Within the context of internal displacement, there are four identifiable links between climate change and internal displacement, namely: (a) sudden on-set disasters; (b) slow on-set disasters; (c) climate-related conflicts; and (d) climate development-induced displacement.

The nexus between technology and internal displacement is an issue for which significant research, which is imperative, is yet to emerge. Given the growing impact of technology and the fact that the digital age is acknowledged both as the present and future, developing appropriate solutions to the challenge of internal displacement in this context is imperative. Within this context, this chapter locates itself, discussing the importance of technology as an IDP research agenda and the

---

[96] Human Rights Watch (2005), Masava (2012), Obiadi et al. (2019, 50).

[97] Parts of this issue has been discussed in this chapter under the section on climate change.

imperative of advancing the conversation on protecting IDPs within this emerging context. While the growing presence of technology presents an important opportunity for developing significant solutions to global challenges, it is imperative to consider technology in IDP research to advance sustainable solutions to the various dimensions of internal displacement, particularly in emerging global narratives. Although quantitative studies establishing the link more firmly is crucial, this chapter identifies two dimensions: technology-enabled attacks and technology-enabled generalized violence.

The negative attitudinal orientation towards foreign nationals has continued to test the moral firmaments of nation-states globally. In Africa, this has emerged in violence and exclusionary policies that reinforce binary relations within democratic societies. While the narrative on xenophobia within the migration discourse has examined issues of criminalisation of migration and mass expulsion, a significant dimension of this narrative that has not received as much attention is internal displacement. This chapter identifies two dimensions of this rhetoric: xenophobia-induced generalized violence and the implementation of discriminatory policies against specific groups of foreign nationals. However, more quantitative studies are required to further highlight the various dimensions of xenophobia and internal displacement and to spotlight other regions where such situations exist in Africa.

Harmful practices are a significant root cause of internal displacement in Africa. However, not much attention has been given to this issue within the IDP discussion. This chapter examines this nexus. However, it is pertinent to reinforce the fact that more studies are required to engage with this interlinkage and also draw connections with other harmful practices causing internal displacement in Africa. This chapter identifies two dimensions of internal displacement from harmful practices relating to gender-based and group-based practices. While specific issues are highlighted in this regard, particularly the issues of sexual slavery, breast ironing female genital mutilation and child marriage, it is pertinent that the nexus between internal displacement and other forms of gender-based practices are also examined. Moreover, within the context of group-based practices specific attention also needs to be paid to other pertinent groups including persons with disabilities and older persons in addition to persons living with albinism.

The nexus between generalized violence and internal displacement in Africa has become a pressing concern for which sustainable solutions are required in the context of internal displacement. Given the prevalence of this root cause of internal displacement and limited focus on it in the IDP research agenda, spotlighting this issue and understanding the nature in which it resonates provides a lens through to advance adequate response. This chapter identifies five dimensions of this nexus: gang violence, electoral violence, interethnic clashes and violent extremism and technology-enabled generalized violence. While this chapter provides knowledge on this subject, it imperative to advance data within national contexts given its prevalence.

Development projects are significant root causes of internal displacement in Africa. While the economic importance of these projects makes them desirable within various national contexts, the fact that they occasion significant population displacement raises a daunting challenge. Across Africa, this has become an evident reality with the increase in these projects since the early periods of independence, often at a human cost within various societies. While the continent is replete with examples of the nexus between development projects and internal displacement, not much attention has been given to this subject in the literature on the issue. This chapter identifies five notable examples: dams, natural resource extraction, agricultural investment, urban renewal and climate-related projects. However, it is imperative that more research emerges within various national contexts regarding the increasing nexus between these projects and internal population displacement across Africa.

References

Abdi C (2012) Eastleigh bombings, terror dynamics and Kenya's Somali intervention. Al Jazeera 13 October

Adeola R (2021) Development-induced displacement and human rights in Africa: the Kampala Convention. Routledge, London

Adrien SM (2007) The DRC case study: the impact of the "Carbon Sinks of Ibi-Batéké Project on the indigenous pygmies of the Democratic Republic of Congo". In: Abhainn M, Bernard KM, Grey S (eds) Indigenous peoples and climate change: vulnerabilities, adaptation and responses to mechanisms of the Kyoto Protocol: a collection of case studies. The International Alliance of Indigenous and Tribal Peoples of the Tropical Forests, Nairobi

Akonor A (2019) Trokosi: young girls and women still in captivity years after law abolishing practice. Ghana Business News 28 October

Al Jazeera (2019) Gangs kill dozens in series of attacks in northern Nigeria. 10 June

Alindogan J (2018) Nigeria's Fulani-farmer conflict displaces many. Al Jazeera 13 May

Almond K (2017) They escaped child marriage. Now they're speaking out. CNN 22 December

Aluko O (2018) 130,000 people displaced by herdsmen, farmers clashes – Red Cross. The Punch 27 January

Amnesty International (2018a) Families torn apart: forced eviction of indigenous people in Embobut forest, Kenya (AFR 32/8340/2018)

Amnesty International (2018b) Harvest of death: three years of bloody clashes between farmers and herders in Nigeria (AFR/9503/2018)

Amusan L, Abegunde O, Temitope E, Akinyemi TE (2017) Climate change, pastoral migration, resource governance and security: the Grazing Bill solution to farmer-herder conflict in Nigeria. Environ Econ 8(3):35–45

Batterbury S, Ndi F (2018) Land-grabbing in Africa. In: Binns T, Lynch K, Nel E (eds) The Routledge handbook of African development. Routledge, London

Bax P, Sguazzin A, Vecchiatto P (2019) Rising Cape Town gang violence is yet another legacy of apartheid. Bloomberg 24 July

BBC (2010) Ivory Coast: Gbagbo under pressure to stand down. 17 December

BBC (2012) Kenya orders Somali refugees to go to Dadaab. 18 December

BBC (2016) The Kenyan girls who fled their families to escape FGM. 14 July

BBC (2017) Mozambique police warns bald men after ritual attack. 7 June

BBC (2019) Tanzania's FGM safe house for girls in danger of cutting. 26 February

Beymer-Ferris BA, Bassett TJ (2012) The REDD menace: resurgent protectionism in Tanzania's mangrove forests. Glob Environ Chang 22:332–341

Bilyeu AS (1999) Trokosi – the practice of sexual slavery in Ghana: religious and cultural freedom vs. human rights. Indiana Int Comp Law Rev 9(2):457–504

Boaten AB (2001) The Trokosi system in Ghana: discrimination against women and children. In: Rwomire A (ed) African women and children: crisis and response. Praeger, Connecticut

Bradley T (2019) Breast ironing is a harmful practice to young girls that isn't getting sufficient attention. Quartz Africa 8 May

Brookes P (2019) The growing Iranian. Unmanned combat aerial vehicle threat needs U.S. action. Center for National Defense No 3437 (17 September)

Bryne C (2018) Climate change and human migration. Univ Calif Irvine Law Rev 8:761

Burke J (2019a) South African army sent into townships to curb gang violence. The Guardian 19 July

Burke J (2019b) "We are a target": wave of xenophobic attacks sweeps Johannesburg. The Guardian 10 September

Cabot C (2015) Climate change, security risks and conflict reduction in Africa: a case study of farmer-herder conflicts over natural resources in Cote d'Ivoire, Ghana and Burkina Faso 1960–2000. Springer, Cham

Casas CM, Codina B, Redaño A, Lorente J (2004) A methodology to classify extreme rainfall events in the western Mediterranean area. Theor Appl Climatol 77(3–4):139–150

Cernea M (1996) Understanding and preventing impoverishment from displacement: reflections on the state of knowledge. In: McDowell C (ed) Understanding impoverishment: the consequences of development-induced displacement. Berghahn Books, Oxford

Chaskalson R (2017) Do immigrants "steal" jobs in South Africa? What the data tells us. Ground Up 18 September

CNN (2009) Report: scores of albinos in hiding after attacks. 29 November

Cotula L (2013) The great African land grab? Agricultural investments and the global food system. Zed Books, London

Crush J, Tawodzera G, Chikanda A, Tevera D (2017a) Living with xenophobia: Zimbabwean informal enterprise in South Africa (SAMP Migration Policy Series)

Crush J, Tawodzera G, Chikanda A, Ramachandran S, Tevera D (2017b) Migrants in countries in crisis (South Africa case study): the double crisis – mass migration from Zimbabwe and xenophobic violence in South Africa. International Centre for Migration Policy Development

de Greef K (2019) As gang murders surge, South Africa sends army to Cape Town, and the city cheers. The New York Times 13 August

de Wet C (2006) Risk, complexity and local initiatives in forced resettlement outcomes. In: de Wet C (ed) Development-induced displacements: problems, policies and people. Berghahn Books, Oxford

DeGeorges A, Reilly BK (2006) Dams and large scale irrigation on the Senegal river: impacts on man and the environment. Int J Environ Stud 63(5):633–644

Department of Refugee Affairs (2012) Press statement. 13 December. https://www.hrw.org/sites/default/files/related_material/Department%20of%20Refugee%20Affairs%20press%20statement%2013%20December%202012.pdf

Dixon R (2017) In parts of Africa, people with albinism are hunted for their body parts. The latest victim: a 9-year-old boy. Los Angeles Times 15 June

Downing T, Garcia-Downing C (2009) Routine and dissonant cultures: a theory about the psycho-socio-cultural disruptions of involuntary displacement and the ways to mitigate them without inflicting even more damage. In: Oliver-Smith A (ed) Development and dispossession: the anthropology of displacement and resettlement. School for Advanced Research Press, Santa Fe

Ennaji M (2019) Child marriage in North Africa: still a lot to be done. The Conversation 23 October

Equality Now (2019) FGM and the law around the world. 19 June

Fahim HM (1981) Dams, people and development: the Aswan High Dam Case. Pergamon Press, Oxford

Farah S (2001) Taking note of human rights. Christ Sci Monit 13 February

Fasona MJ, Omojola AS (2005) Climate change, human security and communal clashes in Nigeria. Paper presented at the international conference on human security and climate change, Asker, Norway, 21–23 June

Fem E (2011) In Cameroon, a conflict rages over breast ironing. Real Change (News) 3 June

Female Genital Mutilation (FGM). https://www.actionaid.org.uk/about-us/what-we-do/violence-against-women-and-girls/female-genital-mutilation-what-is-fgm

FIAN (2012) Land grabbing in Uganda: evictions for foreign investment in coffee in Mubende. July

FIAN International (2012) Coffee to go – with a taste of eviction. A Film from: Michael Enger. https://www.youtube.com/watch?v=InjgvSAcafc

FIGO News (2018) My experience with FGM. 4 October

Freeman L (2017) Environmental change, migration, and conflict in Africa: a critical examination of the interconnections. J Environ Dev 26(4):351–374

Gardner T (2019a) Go and we die, stay and we starve": the Ethiopians facing a deadly dilemma. The Guardian 15 May

Gardner T (2019b) Shadow falls over Ethiopia reforms as warning of crisis go unheeded. The Guardian 14 March

Gausset Q, Whyte M (2012) Climate change and land grab in Africa: resilience for whom? In: Hastrup K, Olwig KF (eds) Climate change and human mobility: global challenges to the social sciences. Cambridge University Press, Cambridge

Gettleman J (2006) Chased by gang violence, residents flee Kenyan slum. The New York Times 10 November

Gillard LM (2010) Trokosi: slave of the gods. Xulon Press, Californi

Grainger M, Geary K (2011) The New Forests Company and its Uganda plantations. Oxfam 22 September

Hans B (2015) King's anti-foreigner speech causes alarm. IOL 23 March

Hausfather Z (2019) UNEP: 1.5C climate target "slipping out of reach". Carbon Brief 26 November

Hugo G (2011) Lessons from past forced resettlement for climate change migration. Piguet E, Pécoud, A and de, In: Guchteneire P (ed) Migration and climate change. Cambridge University Press, Cambridge

Human Rights First (2013) New encampment policy fuels xenophobia in Kenya. Human Right First 8 February

Human Rights Watch (2005) Zimbabwe: evicted and forsaken: internally displaced persons in the aftermath of Operation Marambatsvina. Human Rights Watch, New York

Human Rights Watch (2008) Ballots to bullets: organized political violence and Kenya's crisis of governance. Human Rights Watch, New York

Human Rights Watch (2012a) Kenya: end security force reprisals in north. Human Rights Watch 22 November

Human Rights Watch (2012b) Criminal reprisals: Kenyan police and military abuses against ethnic Somalis. Human Rights Watch, New York

Human Rights Watch (2013) Kenya: don't force 55,000 refugees into camps. Human Rights Watch 21 January

Human Rights Watch (2017) Kenya: post-election killings, abuse. 27 August

Internal Displacement Monitoring Centre (2018) Ethiopia tops global list of highest internal displacement in 2018. 12 September

Internal Displacement Monitoring Centre (2019) Global report on internal displacement. Internal Displacement Monitoring Centre, Geneva

International Crisis Group (2017) Herders against farmers: Nigeria's expanding deadly conflict. Africa Report No 252, 19 September

International Crisis Group (2018) Stopping Nigeria's spiralling farmer-herder violence. Africa Report No 262, 26 July

Jegede AO (2017) Indigenous peoples, climate migration and international human rights law in Africa, with reflections on the relevance of the Kampala Convention. In: Mayer B, Crépeau F (eds) Research handbook on climate change, migration and the law. Edward Elgar, Cheltenham

Jidovanu N, Otieno B (2019) The boarding school in Kenya that helps Maasai girls escape FGM. Al Jazeera 7 March

Kachika T (2010) Land grabbing in Africa: a review of the impacts and the possible policy responses. Oxfam, Nairobi

Kälin W (2010) Conceptualising climate-induced displacement. In: McAdam J (ed) Climate change and displacement: multidisciplinary perspectives. Hart Publishing, London

Kenya Citizen TV (2012) Al-Shabaab targeting security officers. 13 December. https://www.youtube.com/watch?v=vWrFiUnja74

Kerry Kennedy Cuomo. Camera Works: 'Speak truth to power' Kerry Kennedy Cuomo interview with Juliana Dogbadzi. The Washington Post. https://www.washingtonpost.com/wp-srv/photo/onassignment/truth/st/09.htm

Kiptoo R (2016) Kenyan girls hide in schools to escape FGM. BBC (News) 21 December

Klein MA (2014) Historical dictionary of slavery and abolition. Rowman & Littlefield, Maryland

Kleinfeld P (2020) Who's behind the violence in Mozambique's Cabo Delgado. The New Humanitarian 12 February

Koigi B (2017) Protecting Cameroon girls from rape through breast ironing. Fair Planet 6 November

Langat A (2014) On the edge of home: the forcible evictions of the Sengwer in Kenya. Think Africa Press 28 April

Lind J, Mutahi P, Oosterom M (2017) "Killing a mosquito with a hammer": Al-Shabaab violence and state security responses in Kenya. Peacebuilding 5(2):118–135

Lorraine Fiadjoe v Attorney General of the United States, 411 F.3d 135 (3d Cir. 2005)

Lott FC, Christidis N, Stott PA (2013) Can the 2011 East African drought be attributed to human-induced climate change? Geophys Res Lett 40(6):1177–1181

Mabikke S (2014) Large-scale land acquisitions and smallholder farmers in Uganda. In: Nhamo G, Chekwoti C (eds) Land grabs in a green African economy: implications for trade, investment and development policies. Africa Institute of South Africa, Pretoria

Madebwe C, Madebwe V, Mavusa S (2011) Involuntary displacement and resettlement to make way for diamond mining: the case of Chiadzwa villagers in Marange, Zimbabwe. Journal of Research in Peace. Gend Dev 1(10):292

Makochekanwa A (2014) The impacts of foreign land deals on selected African local communities. In: Nhamo G, Chekwoti C (eds) Land grabs in the green African economy: implications for trade, investment and development policies. Africa Institute of South Africa, Pretoria

Masava M (2012) Mitumba slum dwellers evicted, land fenced off. The Star 9 January

McAdam J (2012) Climate change, forced migration and international law. Oxford University Press, Oxford

Mistiaen V (2013) Virgin wives of the fetish gods – Ghana's trokosi tradition. Reuters 4 October

Morris C (2019) The violent work of South African gangs. Alternative Information and Development Centre 8 July

Mushieta O, Merrill A (2010) IBI Bateke Carbon Sink Plantation: an African forestry pilot case. Carbon Clim Law Rev 4(4):351

Nzwili F (2012) After grenade attacks, Kenya wants Somali refugees in camps. The Christian Science Monitor 17 December

Obaji Jr P (2020) "No girl is safe": the mothers ironing their daughters' breasts. Al Jazeera 3 February

Obiadi BN, Onochie AO, Uduak PU (2019) Where is home for the Abuja, Nigeria urban poor. Mgbakoigba J Afr Stud 8(1):50

Obulutsa G (2018) Violence in southern Ethiopia forces more than 800,000 to flee. Reuters (4 July)

Odula T (2012) MSF: more refugees to worsen state of Kenya camp. Associated Press 29 December. https://news.yahoo.com/msf-more-refugees-worsen-state-kenya-camp-144058421.html

Okiror S (2018) The Ugandan girl who trekked barefoot to escape marriage at 13. The Guardian (UK) 26 June

Olaniyan A, Francis M, Okeke-Uzodike U (2015) The cattle are "Ghanaians" but the herders are strangers: farmer-herder conflicts, expulsion policy, and pastoralist question in Agogo, Ghana. Afr Stud Q 15(2):53

Oliver-Smith A (2005) Applied anthropology and development-induced displacement and resettlement. In: Kedia, S and van Willigen, J (eds) Applied anthropology: domains of application. Praeger, Connecticut

Oluyemi F (2014) Displacements in the context of social crises in the oil-rich Niger-Delta of Nigeria and oil-rich Bakassi Peninsula in Cameroon. Int J Soc Work Hum Serv Pract 2(1):28

Opiyo P, Githinji P (2011) Minister says Eastleigh the next target in hunt for Al Shabaab militants. Standard Media 20 October

Paris Agreement (2015) https://unfccc.int/sites/default/files/english_paris_agreement.pdf

Pemunta NV (2016) The social context of breast ironing in Cameroon. Athens J Health 3(4):33

Penz P, Drydyk J, Bose PS (2011) Displacement by development: ethics, rights and responsibilities. Cambridge University Press, Cambridge

Provincial Government of KwaZulu-Natal (2015) Report of the Special Reference Group on Migration and Community Integration in KwaZulu-Natal, 31 October

Reini J (2013) Kenya tensions spark Somali refugee flight. Al Jazeera 8 February

Relocation of urban refugees to officially designated camps (2013) Letter from E Mutea Iringo, Permanent Secretary, Provincial Administration and Internal Security, Office of the President, Kenya to Mr. Andrew A.A. Mondoh, Permanent Secretary, Ministry of Special Programmes, Nairobi, 16 January. https://www.hrw.org/sites/default/files/related_material/16%20January%202013%20letter%20from%20Ministry%20of%20Public%20Administration%20and%20Internal%20Security.pdf

Reuveny R (2007) Climate change-induced migration and violent conflict. Polit Geogr 26:656–673

Rice X (2010) Conflict looms over Ivory Coast while poll-loser Gbagbo refuses to cede control. The Guardian 6 December

Rinelli L, Opondo SO (2015) Affective economies: Eastleigh's metalogistics, urban anxieties and the mapping of diasporic city life. In: Demissie F (ed) Africans on the move: migration, diaspora and development nexus. Routledge, London

Sanghani R (2015) Meet the amazing woman running a safe house for girls fleeing FGM. The Telegraph 2 April

Schmitd-Soltau K (2006) Indigenous peoples planning framework for the western Kenya community driven development and flood mitigation project and the natural resource management project. Republic of Kenya, Nairobi

Scudder T (2005) The future of large dams: dealing with social, environmental, institutional and political costs. Earthscan, London

Seepersaud M (2017) The plight of Tanzanian persons with albinism: a case for international refugee and asylum-procedure reform. Emory Int Law Rev 32:115

Selby D, Ngalle J (2018) The sad reason African mothers 'iron': their daughter's breasts. Global Citizen 6 April

Sengupta S (2018) Hotter, drier, hungrier: how global warming punishes the world's poorest. The New York Times 12 March

South African History Online (2015) Xenophobic violence in democratic South Africa. https://www.sahistory.org.za/article/xenophobic-violence-democratic-south-africa

Sperber A (2019) U.S. Bombardments are driving Somalis from their homes. Foreign Policy 7 March

Straziuso J (2013) Somalia's 2011 famine a consequence of human-induced climate change: study. Vancouver Sun 15 March

Sylla MB, Nikiema PM, Gibba P, Kebe I, Klutse NAB (2016) Climate change over West Africa: recent trends and future projections. In: Yaro JA, Hesselberg J (eds) Adaptation to climate change and variability in rural West Africa. Springer, Cham

Tamakloe MA (1994) Long-term impacts of resettlement: the Akosombo dam experience. In: Cook CC (ed) Involuntary resettlement in Africa: selected papers from a conference on environment and settlement issues in Africa. World Bank Publications, Washington

Tetchiada S (2006) Cameroon: "breast-ironing" of young girls, a harmful custom. 13 June. http://www.wluml.org/node/4129

Thamae ML (2006) A decade of advocacy for dam-affected communities. In: Thamae ML, Pottinger L (eds) On the wrong side of development: lessons learned from the Lesotho highlands water project. Transformation Resource Centre, Maseru

The Guardian (2018) With gangs aplenty, Lagos inner streets know little joy. 13 November

The New Humanitarian (2001) Burundi situation not conducive for refugee return – UNHCR. 29 May

The News Humanitarian (2009) Funds needed for displaced Zimbabweans. 18 November

UN Commission on Science and Technology for Development (2019) The impact of rapid technological change on sustainable development. Report of the Secretary-General. 4 March

UN Environment Programme (2019) Emissions gap report 2019

UN High Commissioner for Human Rights (2008) Report from OHCHR fact-finding mission to Kenya, 6–28 February

UN Human Rights Council (2014) Report of the special rapporteur on the promotion and protection of human rights and fundamental freedoms while countering terrorism, Ben Emmerson, UN Doc A/HRC/25/59, 11 March

UN Human Rights Council (2017) Report of the Independent Expert on the enjoyment of human rights by persons with albinism on her mission to the United Republic of Tanzania, UN Doc A/HRC/37/57/Add. 1, 20 December

UN Office for the Coordination of Humanitarian Affairs (2011) Electoral violence and displacement. 25 March. https://www.refworld.org/pdfid/4d90296a2.pdf

UN Women (2019) Survivors speak: Women leading the movement to end FGM. 4 February

UNHCR (2009) South Africa: UNHCR condemns xenophobic violence in Western Cape. 20 November

UNHCR (2019) Amid rising xenophobic attacks in South Africa, UNHCR ramps up aid for refugees, calls for urgent action. 20 September

UNHCR (n.d.). Nigeria emergency: the Boko Haram insurgency has displaced nearly 2.4 million people in the Lake Chad basin. https://www.unhcr.org/nigeria-emergency.html

UNICEF (2018) Achieving key results for children in Niger (KR4C # 6): ending child marriage. Niger Issue Brief, May

Wambua-Soi, C (2012) Rising xenophobia against Somalis in Kenya. Al Jazeera 20 November

West African Democracy Radio (2017) Guinea: South Africa's Anglo-Gold Ashanti accused of wrongful eviction of villagers. 2 February

Wodon Q, Male C, Montenegro C, Nguyen H, Onagoruwa A (2018) The cost of not educating girls: educating girls and ending child marriage: a priority for Africa. World Bank, Washington, DC

World Bank (2007) Project appraisal document on a proposed credit in the amount of SDR 46.0 million (US$68.5 million equivalent) to the government of Kenya for a natural resource management project. Report no. 37982-KE, 26 February

World Bank (2009) Democratic Republic of Congo – Ibi Bateke Carbon Sink Plantation Project http://documents.worldbank.org/curated/en/456281468235145627/pdf/487450PID0Congo0Ibi0Bateke0Box338924.pdf

World Bank (2018a) Africa loses billions of dollars due to child marriage, says new World Bank. 20 November

World Bank (2018b) Groundswell: preparing for internal climate migration. World Bank, Washington, DC

World Bank Inspection Panel (2013) Report and recommendation – Kenya: natural resource management project (P095050). Report No. 77959-KE, 29 May

Xinhua News Agency (2001) Tanzania orders refugees to go back to camps. 21 May

Yarnell M (2014) Kenyan government cracking down on Somali refugees. Refugees International 11 July

Yarnell M, Cone D (2019) Devastation and displacement: unprecedented cyclones in Mozambique and Zimbabwe a sign of what's to come?' Refugee International 13 August

Zahiri EP, Bamba I, Famien AM, Koffi AK, Ochou AD (2016) Mesoscale extreme rainfall events in West Africa: the cases of Niamey (Niger) and the upper Ouémé valley (Benin). Weather Climate Extremes 13:15–34

Zounmenou DD, Lamin AR (2011) Côte d'Ivoire's post-election crisis: Ouattara rules but can he govern? J Afr Elections 10(2):6–21

Chapter 4
Legal Protection in Emerging Contexts

Abstract In the previous chapter, the emerging root causes of internal displacement were examined with a view to understanding the interlinkages between these issues and situations of internal displacement. Having examined these issues, it relevant to turn to the next area of focus which is how these issues may be addressed from a legal standpoint. This chapter builds on the premise that protecting internally displaced persons (IDPs) within these emerging contexts, from a legal standpoint, requires an understanding of the legal corpus on IDP protection and its application within emerging contexts. In advancing the discussion, this chapter begins with a discussion of the normative frameworks on internal displacement, providing an exposition on these frameworks, first from an historical background to ground an understanding of the internal displacement corpus. This chapter proceeds to discuss the scope of protection within these frameworks from a comparative perspective, analysing the provisions and drawing out the central theme of these frameworks: the right not to be arbitrarily displaced. Afterwards, this chapter considers the protection for IDPs in emerging contexts drawing on the implication of the right not to be arbitrarily displaced and what it entails for the protection of persons within the various contexts.

Keywords Internally displaced persons · Internal displacement · Emerging contexts · Legal protection · UN Guiding Principles · Kampala Convention · Great Lakes Protocol

4.1 Process

This section discusses the normative framework on internal displacement focusing on the development of the frameworks relevant to the legal corpus on IDPs along three main areas: global, regional and sub-regional levels in Africa.[1]

---

[1] UN Commission on Human Rights, Addendum, 'Guiding Principles on Internal Displacement' *Report of the Representative of the Secretary-General, Mr. Francis M. Deng, submitted pursuant to Commission on Human Rights resolution 1997/39,* UN Doc. E/CN.4/1998/53/Add.2 (11 February 1998) (Guiding Principles), International Conference on the Great Lakes Region *Protocol*

© The Author(s), under exclusive license to Springer Nature Switzerland AG 2021
R. Adeola, *Emerging Issues in Internal Displacement in Africa*, SpringerBriefs in Law, https://doi.org/10.1007/978-3-030-64562-5_4

4.1.1   UN Guiding Principles on Internal Displacement

The global framework on internal displacement emerged in response to the normative gap in the protection of persons displaced internally within state borders in need of protection and assistance having fled from their homes and places of habitual residence.[2] However, the development of the norm for the protection of IDPs was never without its challenges. During early discussions on the protection of these persons, there were some concerns that developing IDP norms will recast focus on in-country protection and receiving states will leverage on this to 'preclude their asylum obligations' under international refugee law.[3] Given that those displaced remain within the state of displacement, there was also the concern that a global regulatory framework may become a pretext for eroding state sovereignty.

However, with the global shift towards sovereignty as responsibility,[4] states are increasingly aware of the need to ensure a responsible outlook in how issues of protection for displaced populations are managed. And the global framework on which protection is curated is the UN Guiding Principles on Internal Displacement. Notable mention of the need for a framework to respond to the protection needs of IDPs was mooted at the International Conference on the Plight of Refugees, Returnees and Displaced Persons in Southern Africa (SARRED) held in Norway in 1988.[5] At SARRED, it was resolved that a study should be conducted and deliberations held to determine whether there was a need for a UN arrangement for the benefit of IDPs.[6] Upon the request of the UN Commission on Human Rights,[7] the UN Secretary-General subsequently appointed Francis Deng as Special

---

on the Protection and Assistance of Internally Displaced Persons in the Great Lakes Region (2006), African Union Convention for the Protection and Assistance of Internally Displaced Persons in Africa, adopted at the Special Summit of the African Union Heads of States and Government in Kampala, Uganda (19–23 October 2009) (the Kampala Convention).

[2] See Cohen and Deng (1998, 76), Phuong (2004, 53–56), Kälin (2014a, 612–633), Cantor (2018), Cardona-Fox (2019), Orchard (2019), Adeola (2020).

[3] Phuong (2004, 4).

[4] Evans (2015, 16–37), Genser and Cotler (2012), Kuwali and Viljoen (2014), Bellamy (2015), Breau (2016).

[5] UN General Assembly (19 October 1988a).

[6] Paragraph 21 of the Oslo Declaration was to the effect that: [i]n view of the absence of a United Nations operational body specifically charged to deal with the problems of and assistance to internally displaced persons, the Secretary-General of the United Nations is requested to undertake studies and consultations in order to ensure tile timely implementation and overall co-ordination of relief programmes for these people. *Oslo declaration and plan of action on the plight of refugees, returnees and displaced persons in southern Africa* 1988, para 21; This was supported by UN General Assembly Resolution 43/116 that mandated the UN Secretary-General to 'undertake studies and consultations in order to consider the need for the establishment, within the United Nations system, of a mechanism or arrangement to ensure the implementation and overall co-ordination of relief programmes to internally displaced persons'. UN General Assembly (8 December 1988b), para 6.

[7] UN Commission on Human Rights (1992).

Representative on the issue of internal displacement – to examine the existing gaps in international law and develop recommendations on how best to protect IDPs.[8]

Existing protection at the time, relevant to the protection of displaced persons, was codified in three main instruments: international humanitarian law, human rights law and refugee law. The fact that IDPs were within the country of displacement made refugee law inapplicable. Although, comparatively, it could serve as a benchmark for standard-setting also for IDPs given that the UN 1951 Refugee Convention was detailed on the protection of refugees and provided notable norms worth considering for the protection of IDPs. However, applying the UN 1951 Refugee Convention to IDPs was difficult given the nature of its coverage. On its part, international humanitarian law was of much relevance. However, its holistic relevance to the issue of internal displacement was also affected by the fact that it applied in situations of armed conflict, which begs the question as to what happens in situations of generalized violence or other circumstances that fall short of the threshold of armed conflict. Moreover, it was not also applicable in situations of peace, unlike human rights, which was a pertinent short-coming especially where the root cause of displacement was not conflict. In this regard, human rights norms were pertinent and of much importance to the protection of IDPs. However, these human rights norms also had their shortcomings, one of which was that they could be limited. Moreover, the standard of limitation of rights especially in the context of internal displacement was not specifically defined. Although these norms were of much significance, having a specific framework for the protection of IDPs was crucial. However, it was not clear in the 1990s how this protection should be actualised. What was clear, nonetheless, was the fact that there was a need for global action. Following the SARRED Conference, this was recognised by the UN, states and international agencies.

But having a binding framework was a less desirable option. First, there were concerns of state sovereignty which may have stalled or stunted the process. In addition, it seemed also a more plausible and strategic option given that the development of a binding instrument in the early 1990s would have been subjected to a much vigorous process that may have been counterproductive or even watered-down protection for IDPs. Indeed, the articulation of the Guiding Principles was, at the time, a step in the right direction given that it could later serve as a normative springboard for the evolution of a set of binding norms when there is a critical mass of global consensus among states on the imperative for a legal framework for the protection of IDPs. Indeed, the Guiding Principles has become hortatory inducing a protection-oriented approach in the furtherance of protection for displaced populations.[9] The Guiding Principles have become the parlance on which the actions of states in the protection of IDPs are analysed.

For more than two decades following its development, the global framework has become an imperative locomotive for advancing solutions to the development of

---

[8] Deng (2001, 141).

[9] Groth (2011, 7), Kälin (2001, 6).

norms and institutional responses to the protection of persons displaced within the borders of the state due to a plethora of circumstances. As with the UN 1951 Refugee Convention, the Guiding Principles is governed by the impetus of affording adequate protection to displaced populations and setting out pertinent obligations in this respect. However, unlike the UN 1951 Refugee Convention, the Guiding Principles are a set of non-binding norms, recognized by UN Member States as 'an important international framework for the protection of internally displaced persons'.[10] Though a soft norm, it has become a notable springboard for hard norms addressing the plight of IDPs. In addition, the Guiding Principles respond to the vulnerability and deprivations that persons who have been forced or compelled to flee but have not crossed internationally recognized borders may experience, including violence, forced returns, deprivation of humanitarian assistance and forcible resettlement.[11] It emphasises the importance of national protection and further stresses the importance of cooperation in reaching durable solutions.

4.1.2   AU Convention on Internally Displaced Persons

With the existence of the Guiding Principles at the global level, there is a question to be answered as to why the African Union (AU) decided to develop a binding framework on internal displacement given that there was also already an existing norm at the global level for the protection of IDPs. Similar questions may be raised for the development of other regional norms that seek to replicate global protection such as the human rights framework and even the norm on the protection of refugees. There is a plethora of answers for this question. But there are five pertinent responses that are worth considering.

The first relates to the notion of ownership that is often reflected in the preamble of several treaties as a motivation for the development of a regional norm. Under the African Charter on Human and Peoples' Rights for instance, AU Member States were mindful of 'historical tradition' and sought the 'values of African civilization' which was to serve as an inspiration for 'the concept of human and peoples' rights'.[12] And indeed, in the title and content of the treaty, these virtues are expressed in the notions of people's rights, in the concept of duties and significantly in the provision of the right to development. Notably, also, under the OAU 1969 Refugee Convention, states expressed a collective conviction that 'all the problems of our continent must be solved in the spirit of the Charter of the Organization of African Unity and in the African context'.[13] The Kampala Convention was no different in this pattern of rea-

---

[10] World Summit Outcome (2005, para 132).

[11] Kleine-Ahlbrandt (2004, 16).

[12] *African Charter on Human and Peoples' Rights* (1981), para 4 of the preamble.

[13] See *OAU Convention Governing the Specific Aspects of Refugee Problems in Africa* (1969), para 8 of the preamble.

soning given that it also provides that it seeks to reiterate 'inherent African custom and tradition of hospitality by local host communities for persons in distress and support for such communities'.[14]

The development of regional frameworks can also be seen as an avenue for proclaiming support or objecting to global processes. A notable example of the latter, for instance, is with the adoption of an AU Protocol expanding the jurisdiction of the proposed regional court of justice and human rights to criminal matters on the heels of a perceived global prejudice to the fight against criminal impunity.[15] As a retort to the global criminal justice pendulum, the Malabo Protocol recognises immunity for heads of state and government in criminal justice administration at variance with the Rome Statute of the International Criminal Court.[16] However, it is rarely the case that normative frameworks are developed as objections to global norms and processes. Often, normative frameworks are used to lend support and cement collective recognition of global normative frameworks. In the Kampala Convention, African leaders emphasise this in recognising the rights of IDPs 'as provided for and protected in international human rights and humanitarian law and as set out in the 1998 United Nations Guiding Principles on Internal Displacement'.[17] Closely, it also seeks to mirror a plethora of other instruments at the global level including the Universal Declaration of Human Rights and the Four Geneva Conventions.

Moreover, there is also a subtle motivation mostly reflected in legal scholarship of a need for African voices in international law. One might argue that this motivation branches out of the critical mass of scholarship of proponents of Third World Approaches to International Law – a critical field of scholarship where scholars pose questions on the legitimacy of the content of international law and processes of its formation.[18] Proponents of African approaches to international law emphasise the importance of infusing African values into the international legal discourse as a paradigm shift from the global knowledge formations to curate reception at the local level and also build consensus among states and other relevant stakeholders. In this field, African norm-setting is legitimate in substance and process.

Further, the development of the norms is often against the backdrop of breathing life to historical commitments. Aside from the Kampala Convention, a pertinent example of this is in the development of the AU Protocol on Free Movement of Persons, Right of Residence and Right of Establishment in Africa (Free Movement Protocol),[19] which was developed as a means of fulfilling the commitment under the Treaty Establishing the African Economic Community, made under article 43(2) to

---

[14] *Kampala Convention*, para 3 of the preamble.

[15] *AU Protocol on Amendments to the Protocol on the Statute of the African Court of Justice and Human Rights* (2014).

[16] *Rome Statute of the International Criminal Court* (1998).

[17] *Kampala Convention*, para 10 of the preamble.

[18] Mutua (2000, 31–39), Anghie (2004), Okafor (2008, 371).

[19] *Protocol on Free Movement of Persons, Right of Residence and Right of Establishment in Africa, adopted by the AU Heads of States and Government Thirtieth Ordinary Session in Addis Ababa, Ethiopia* (2018).

the effect that AU Member States will adopt 'necessary measures, in order to achieve progressively the free movement of persons, and to ensure the enjoyment of the right of residence and the right of establishment by their nationals'.[20] In the Protocol to the African Charter on the Rights of Older Persons in Africa, a similar commitment resonates.[21] Under the Kampala Convention, states sough to reaffirm a historical commitment to the protection and assistance of IDPs through the development of a distinct legal instrument and this is emphasised with reference to the AU Executive Council decisions of 2004.[22]

Moreover, there is also the rationale that solutions to issues of commonality is effectively addressed on the heels of regional solidarity elevated on the wings of Pan-Africanism. After all, this was why African states established the Organisation for African Unity (OAU). And since this ideology has proved effective in steering actions on institutional development, it seemed plausible to harness its dividend towards common priorities through the avenue of norm-setting. This motivation exudes from one of the primary objectives of the Kampala Convention which is to '[e]stablish a legal framework for solidarity, cooperation, promotion of durable solutions and mutual support between the State Parties in order to combat displacement and address its consequences'.[23] However, it may be argued that an alternative method for 'advancing solidarity, cooperation, promotion of durable solutions and mutual support' could have been to adopt a regional policy implementing the Guiding Principles. But the AU executive organ decided to develop a separate binding framework recognising the fact that there was no separate framework on the protection and assistance of IDPs.[24] There was a legal instrument on refugees which it decided should be retained in its present form and complemented 'through soft norms; through the adoption of annual decisions of the Assembly.'[25] However, on IDPs, the AU Executive Council requested the AU Commission 'to collaborate with relevant cooperating partners and other stakeholders to ensure that Internally Displaced Persons are provided with an appropriate legal framework to ensure their adequate protection and assistance'.[26]

---

[20] *Treaty Establishing the African Economic Community, adopted by the Heads of State and Government of Member States of the Organisation of African Unity in Abuja, Nigeria* (1981), art 43(2).

[21] *Protocol to the African Charter on Human and Peoples' Rights on the Rights of Older Persons in Africa* (2016), para 8 of preamble.

[22] *Kampala Convention*, para 14 of the preamble.

[23] *Kampala Convention*, art 2(c).

[24] AU Executive Council *Decision on the situation of refugees, returnees and displaced persons*, AU Doc EX/CL/127 (V); AU Executive Council *Decision on the meeting of experts on the review of OAU/AU treaties*, AU Doc EX.CL/Dec. 129 (V), para 4(i).

[25] AU Executive Council *Decision on the meeting of experts on the review of OAU/AU* treaties, para 4(ii)

[26] AU Executive Council *Decision on the situation of refugees, returnees and displaced persons*, para 8.

The Kampala Convention was adopted in 2009, in response to the 'sheer gravity' of IDP situations on the continent, the 'precarious existence and vulnerability' experienced by IDPs and 'the dimensions of human rights and humanitarian law required to provide for their physical and material protection, including humanitarian assistance.'[27] There was also the problem of rising populations in of protection within state borders. In response, AU Member States decided to develop a uniform regional standard for protecting and assisting these populations. The decision of the AU led to a 5-year drafting process which culminated in the adoption of the Kampala Convention during a Special Summit of AU Heads of State and Government held in Kampala, Uganda in 2009.[28] Over the last decade, the Kampala Convention has led to the development of laws and policies across Africa.[29] Notably, this has emerged in countries such as Niger, Zambia and Somalia.[30] Other countries have developed draft normative frameworks including Nigeria, the Democratic Republic of Congo and South Sudan.[31] Having been ratified by more than half of the 55 AU Member States, the Kampala Convention has become a pertinent regional framework for the furtherance of IDP protection and assistance. Aside from referencing the Guiding Principles as its normative source, the Kampala Convention further reflects key aspects of this global framework including the prohibition of arbitrary displacement, which is the cornerstone of protection in internal displacement. The next section examines the content of this obligation.

4.1.3 ICGLR Internally Displaced Persons Protocol

In 2004, the International Conference of the Great Lakes Region (ICGLR) adopted a Declaration on Peace, Security, Democracy and Development (Declaration).[32] This Declaration was in response to the conflict situation in the Great Lakes region which had significantly affected socio-economic development and progress in the region, dating back to the early period of post-colonial epoch and in certain

---

[27] Beyani (2006, 189).

[28] Abebe (2016).

[29] Adeola (2019).

[30] Zambia: Guidelines for the compensation and resettlement of internally displaced persons (2013), Niger: Law No. 2018–74 of 10 December 2018, relating to the Protection and Assistance of Internally Displaced Persons (2018), Somalia: National Policy on Refugee-Returnees and Internally Displaced Persons (2019).

[31] See Adeola (2020 85–86), Nigeria: Rights of Internally Displaced Persons (IDPs) Bill (2016), Democratic Republic of Congo: Bill providing protection and assistance to internally displaced persons (2014), South Sudan: Protection and Assistance to Internally Displaced Persons Act 2019 (Draft).

[32] *Declaration on Peace, Security, Democracy and Development* (2004), The ICGLR comprises of 12 states: Angola, Burundi, Central African Republic, Democratic Republic of Congo, Kenya, Republic of Congo, Rwanda, South Sudan, Sudan, Uganda, Tanzania and Zambia.

instances, prior to this time. In the Declaration, states agreed to leverage on the Guiding Principles in the furtherance of protection and assistance to IDPs.[33]

In 2006, a Pact on Security, Stability and Development in the Great Lakes Region (the Pact) was adopted to provide a comprehensive approach to the objectives of states in the Great Lakes Region towards sustainable peace and resolution of conflict situations in the region.[34] The Pact was to reinforce the Declaration earlier adopted in 2004 in which the commitment towards securing lasting peace and fostering a collective front towards the establishment of stability was reinforced.[35] Under the Pact, states commit towards the development of a Protocol on IDPs in line with the Guiding Principles.[36]

In arriving at this objective, the Great Lakes Protocol was adopted in 2006.[37] Through seven provisions, the Great Lakes Protocol reinforces the importance of the Guiding Principles as the regional optics for the protection and assistance to IDPs.[38] In line with article 2 of the Great Lakes Protocol, states agree to develop 'a legal framework in the Great Lakes Region for ensuring the adoption and implementation by Member States of the Guiding Principles on Internal Displacement'.[39] Annexed to this instrument are the Guiding Principles and a Model legislation for the development of normative guidance on the protection and assistance of IDPs. Within the Great Lakes Region, Kenya developed a national framework leveraging on the Great Lakes Protocol and reinforcing the provisions of the Guiding Principles within the national context.[40] Emphatically, the Great Lakes Protocol reinforces states' obligation to 'prevent arbitrary displacement and to eliminate the root causes of displacement.'[41]

---

[33] *Declaration on Peace, Security, Democracy and Development* (2004, para 58).

[34] *Pact on Security, Stability and Development in the Great Lakes Region* (2006), Beyani (2007, 173).

[35] *Pact on Security, Stability and Development in the Great Lakes Region* (2006).

[36] Article 12 of the Pact on Security, Stability and Development in the Great Lakes Region provides that 'The Member States undertake, in accordance with the Protocol on the Protection and Assistance to Internally Displaced Persons, to provide special protection and assistance to internally displaced persons and in particular to adopt and implement the Guiding Principles on Internal Displacement as proposed by the United Nations Secretariat.' *Pact on Security, Stability and Development in the Great Lakes Region* (2006).

[37] International Conference on the Great Lakes Region *Protocol on the Protection and Assistance of Internally Displaced Persons* (2006), Pact on Security, Stability and Development in the Great Lakes Region (2006), Kälin (2014b, 163).

[38] Bernstein and Bueno (2007, 73).

[39] International Conference on the Great Lakes Region *Protocol on the Protection and Assistance of Internally Displaced Persons in the Great Lakes Region* (2006), art 2.

[40] Prevention, Protection and Assistance to Internally Displaced Persons and Affected Communities Act (2012).

[41] International Conference on the Great Lakes Region *Protocol on the Protection and Assistance of Internally Displaced Persons* (2006), art 3(1).

4.2   Content

It is relevant to point out that the global standard on internal displacement is soft law, what this implies is that it is non-binding and cannot be ratified by states. In the early 1990s, when the development of a framework on internal displacement was being discussed, a comprehensive approach through soft law was preferred since negotiating a treaty will have encountered resistance as IDPs were within the borders of states and not having crossed international boundaries, their protection and assistance was regarded as being within the internal affairs of the state of displacement.

Soft laws in the international legal context have become a prominent feature. While not recognised as a conventional source of international law in terms of article 38(1) of the Statute of the International Court of Justice,[42] these laws are often preferred choices in resolving contentious global issues, for instance, on climate change and migration. Guzman and Meyer argue that these forms of laws are used for four pertinent reasons: as a way of coordinating compliance through a focal point (coordination theory), as a means of avoiding the adverse effect of non-compliance with legal rules or sanction (loss avoidance theory); as an avenue to control the outcome regarding the desirability of a particular legal rule (delegation theory); and as a way around consent to a particular issue by assigning soft law-making powers to non-state entities – what they describe as international common law (ICL) theory.[43] With the Guiding Principles, a soft law approach was chosen early on due to the rigours of treaty negotiation that may not only be time consuming but futile particularly in a situation where the avoidance of sanctions was pragmatic. Over the last decades, the Guiding Principles has gained traction with the emergence of binding treaty law in Africa and through the development of national laws across various contexts. Though it is soft law, its influence has lent support to the argument that its provisions are beginning to develop into customary international law. This will require state practice and *opinio juris*. However, the fact that it emerged as an authoritative text in protecting and assisting IDPs reflects its importance even while being soft law.

Much of the provisions of the Kampala Convention and the Great Lakes Protocol draw normative strength from the Guiding Principles. There are four pertinent areas in which this normative strength may be observed: prevention, protection, humanitarian assistance and durable solutions.

With respect to prevention, the Guiding Principles is emphatic on the need for states to ensure respect for their international legal obligations that reinforces the obligation to prevent arbitrary displacement. Both in the Kampala Convention and the Great Lakes Protocol, this is emphasised.[44] These obligations resonate within the context of human rights and humanitarian law. The corpus of international

---

[42] *Statute of the International Court of Justice* (1946), art 38(1).

[43] Guzman and Meyer (2010, 171).

[44] As above, art 4(1)(a).

human rights and humanitarian law incorporate a plethora of frameworks that significantly relate to IDPs from various perspectives. In building on these frameworks, the law on internal displacement essentially seeks to ensure that prevention of internal displacement is the central starting point for realising IDP rights not to be arbitrarily displaced, which represents the corpus of legal protection for IDPs. Drawing on various norms both from thematic and group-based perspectives, prevention is integral to developing a significant response to internal displacement. Much of what is required in addressing internal displacement from the point of root causes, necessarily involves prevention. Generally, it is required that states take measures towards ensuring that internal displacement do not occur. Within the context of various emerging issues such as xenophobia, harmful practices and situations of generalized violence, for instance, understanding prevention as an essential starting point is imperative in avoiding situations of internal displacement altogether. What prevention entails may often intersect with other areas of international law, such as international environmental law, for instance, in the climate change context.

With regards to protection, the Guiding Principles reinforces international human rights and humanitarian law. There are a wide range of pertinent provisions around which this is constructed. In the Guiding Principles, protection draws on the normative provisions relating to the right to life, dignity, liberty and security of persons. The relevance of these provisions resonates from the fact that it places human rights provisions within an understanding of internal displacement and as such reinforces specific areas where attention is required for IDPs within the context of these legal frameworks. The pertinence of this, for instance, emanates from an explicit prohibition of the recruitment or use of children in hostilities under principle 13 of the Guiding Principles.[45] Whereas, under international human rights and humanitarian law, children above 15 may be recruited into the armed forces of a state.[46] Within the

---

[45] Guiding Principles, principle 13.

[46] Article 4(3)(2)(c) of the Protocol Additional to the Geneva Conventions of 12 August 1949 and Relating to the Protection of Victims of Non-international Armed Conflict provides that 'Children who have not attained the age of fifteen years shall neither be recruited in the armed forces or groups nor allowed to take part in hostilities'; similarly, article 38(2) of the Convention on the Rights of the Child require states to 'take all feasible measures to ensure that persons who have not attained the age of fifteen years do not take a direct part in hostilities.' However, article 1 of the Optional Protocol to the Convention on the Rights of the Child on the Involvement of Children in Armed Conflict provides that 'States Parties shall take all feasible measures to ensure that members of their armed forces who have not attained the age of 18 years do not take a direct part in hostilities.' Moreover, the African Charter on the Rights and Welfare of the Child prohibits the recruitment and use of children in armed conflict. Under article 22(2) of the members of their armed forces who have not attained the age of 18 years do not take a direct part in hostilities.' Moreover, the African Charter on the Rights and Welfare of the Child provides that states 'shall take all necessary measures to ensure that no child shall take a direct part in hostilities and refrain in particular, from recruiting any child.' See Protocol Additional to the Geneva Conventions of 12 August 1949 and Relating to the Protection of Victims of Non-international Armed Conflict (1977), art 4(3)(2); Convention on the Rights of the Child (1989), art 38; African Charter on the Rights and Welfare of the Child (1990), art 22(2); Optional Protocol to the Convention on the Rights of the Child on the Involvement of Children in Armed Conflict (2000), art 1.

context of protection, the Guiding Principles further accentuates the right of IDPs to free movement, the right to seek and enjoy asylum, the right to know the fate or whereabouts of their relatives, right to respect for family and the right to an adequate standard of living, including its minimum core in situations of internal displacement which includes: '(a) essential food and potable water; (b) basic shelter and housing; (c) appropriate clothing; and (d) essential medical services and sanitation'.[47] The Guiding Principles further emphasises the protection of specific groups including women, children and pastoral populations. This is reinforced in the Kampala Convention and the Great Lakes Protocol. Moreover, there is an emphasis on the protection of the housing and property rights of IDPs.

In relation to humanitarian assistance, the Guiding Principles emphasises the pertinence of respect for fundamental humanitarian principles, including 'humanity and impartiality'. The Kampala Convention further reinforces the principles of 'neutrality' and 'independence of humanitarian actors'.[48] Both the Guiding Principles and the Kampala Convention are emphatic on the fact that primary responsibility for providing humanitarian assistance is with states.[49] However, where humanitarian actors offer their services in support of IDPs, it is pertinent that states do not arbitrarily withhold their consent regarding such offer or regard this 'as an unfriendly act or an interference in a State's internal affairs and instead, it shall be considered in good faith.'[50] As such, it is crucial that states do not utilise their prerogative to deny protection for IDPs. The Guiding Principles is emphatic that states should not withhold consent 'particularly when authorities concerned are unable or unwilling to provide the required humanitarian assistance.'[51] Moreover, the Great Lakes Protocol emphasises that where states 'lack the capacity' to provide protection and assistance, they 'shall accept and respect the obligation of the organs of the international community to provide protection and assistance to internally displaced persons.'[52] In order to ensure that humanitarian actors can carry out their activities, it is crucial for states to ensure access for these actors by ensuring that the personnel and their facilities are protected and do not form the object of attack.[53] Further, the rights of IDPs to request protection must be protected.

With regards to durable solutions, the Guiding Principles emphasises three pertinent solutions: return, reintegration and resettlement. In arriving at these solutions, the choice of IDPs must be a central theme that guides these solutions. As durable solutions are often long-term in nature, it is crucial to involve IDPs in the planning and management of these solutions. The Guiding Principles further accentuates the

---

[47] *Guiding Principles*, principle 18(2).

[48] *Kampala Convention*, arts 6(3) & 5(8).

[49] *Guiding Principles*, principle 25(1); *Kampala Convention*, art 5(1).

[50] *Guiding Principles*, principle 25(2).

[51] *Guiding Principles*, principle 25(2).

[52] International Conference on the Great Lakes Region *Protocol on the Protection and Assistance of Internally Displaced Persons* (2006), art 3(10).

[53] *Guiding Principles*, principle 26; *Kampala Convention*, art 5(10).

need for states to ensure the participation of IDPs in the provision of durable solutions. This is crucial to the sustainability of solutions also to ensure that a bottom-up approach is adopted, rather than a top-down process that does not significantly incorporate the needs and requirements of IDPs. Where IDPs return, the Guiding Principles reinforces the need for states to ensure that IDPs 'recover, to the extent possible, their property and possessions which they left behind or were dispossessed of upon their displacement.'[54] The Kampala Convention goes further to emphasise the need for states to 'establish mechanisms providing for simplified procedures where necessary, for resolving disputes relating to the property of internally displaced persons.'[55] Overall, states need to ensure the safety of IDPs in the furtherance of durable solutions. Notably, the Great Lakes Protocol reinforces the primary responsibility of states to ensure the safety of IDPs also upon return or resettlement within the state.[56]

Notably, these four pertinent areas find expression in the right not to be arbitrarily displaced which is integral to the protection and assistance of IDPs. This is the primary optic from which the protection of IDPs in emerging contexts is discussed in the next section.

4.2.1   The Right Not to Be Arbitrarily Displaced

Integral to the protection and assistance of IDPs is the right not to be arbitrarily displaced. Both in the Guiding Principles, which the Great Lakes Protocol emphasises and the Kampala Convention, the right not to be arbitrarily displaced is significantly emphasised. Article 4 of the Kampala Convention iterates protection from arbitrary displacement. While the Kampala Convention places an explicit obligation on states to prevent this form of displacement, it does not define what arbitrary displacement constitutes. However, an indication on the notion of arbitrary is presented in article 4(4), specifically, sub-paragraph (h) which recognizes arbitrary displacement a displacement 'caused by any act, event, factor, or phenomenon … which is not justified under international law, including human rights and international humanitarian law.'[57] However, in *Elettronica Sicula S.p.A. (ELSI) (United States of America v Italy)*, the International Court of Justice defines the concept of arbitrary as something that is not just as a situation of unlawfulness but also as 'a willful disregard of due process of law.'[58] Within this context, there are two specific thresholds for understanding arbitrary.

---

[54] *Guiding Principles*, principle 29.

[55] *Kampala Convention*, art 11 (3).

[56] International Conference on the Great Lakes Region *Protocol on the Protection and Assistance of Internally Displaced Persons* (2006), art 3(3)

[57] *Kampala Convention*, art 4(4)(h).

[58] *Elettronica Sicula S.p.A. (ELSI) (United States of America v. Italy)* (Judgement) [1989] ICJ Rep 15, para 128.

The first threshold is legality, or the requirement that an act of internal displacement must be permissible under international law. Within this context, it is useful to examine the root causes examined in this book and whether they are permissible in international law. Under the Kampala Convention, climate change is recognized as a root cause of internal displacement. Linked to natural disasters,[59] displacement may occur in situations of climate change where it is for the safety or health of the population.[60] Situations of technology are not explicitly recognized in the Kampala Convention. However, the Kampala Convention prohibits displacement caused by 'any act, event, factor, or phenomenon of comparable gravity'[61] to other categories of internal displacement recognized in article 4(4). This provision prohibits displacement as a result of generalized violence or violation of human rights. Situations of technology-enabled attacks and technology-enabled generalized violence will generally fall within this context given that they relate to prohibited grounds of internal displacement. Displacement due to harmful practices is explicitly prohibited under the Kampala Convention. Similarly, situations of generalized violence can have no justification as it is also explicitly prohibited. But with respect to development projects, the Kampala Convention permits displacement, although vaguely prescribing the grounds for its permissibility with the words 'as much as possible'.[62] In the Guiding Principles, however, this form of displacement is permissible where there is a 'compelling and overriding public interests'.[63]

A second threshold, however, that must be met is the requirement of due process. There are three ramifications of the due process requirement: general guarantees, cause-specific guarantees and group-specific guarantees. General guarantees include general requirements that must be complied with in the furtherance of protection for IDPs. These guarantees cover areas of protection, including preventing violation of IDPs rights, humanitarian assistance and providing sustainable solutions with the involvement of IDPs in the process. There are also group-specific guarantees in view of the specific provisions of international human right law relating to the protection of groups such as women, children, persons with disabilities, older persons and indigenous peoples.[64] Aside from general and group-specific guarantees, there are cause-specific guarantees that relate to the various root cause of internal displacement. In the context of the issues examined in the previous chapter, these cause-specific guarantees are considered in turn.

---

[59] *Kampala Convention*, art 5(4).

[60] *Kampala Convention*, art 4(4)(f).

[61] *Kampala Convention*, art 4(4)(h).

[62] *Kampala Convention*, art 10.

[63] *Guiding Principles*, principle 6(2)(c).

[64] *Convention on the Elimination of All Forms of Discrimination Against Women* (1979), Convention on the Rights of the Child (n 46), African Charter on the Rights and Welfare of the Child, adopted by the Organisation of African Unity (n 46), United Nations Principles for Older Persons (1991), UN Convention on the Rights of Persons with Disabilities (2006), UN Declaration on the Rights of Indigenous Peoples (2007), Inter-American Convention on Protecting the Human Rights of Older Persons (2015), Protocol to the African Charter on Human and Peoples' Rights on the Rights of Older Persons (2016), Protocol to the African Charter on Human and Peoples' Rights on the Rights of Persons with Disabilities in Africa (2018).

A. Climate Change-Induced Displaced Persons

There are specific due process requirements integral to climate change and internal displacement. First, is the need for adequate planning ahead of climate-related disasters. This requires early warning systems to be established as reinforced in article 4(2) of the Kampala Convention which provides that states 'shall devise early warning systems in the context of the continental early warning system, in areas of potential displacement'.[65] Developing such early warning systems is imperative to advance adequate disaster response and enhance substantial planning for protection of climate change IDPs.

Second, there is a need for engagement with local communities in the process of planning an adequate response. The essence of engagements resonates from the need to be guided by the importance of participation which involves affected populations. Involving these persons in the process will enhance decision-making on prevention, disaster risk management including preparedness, response and planned relocation where required. Moreover, engaging communities will also ensure that an adequate vulnerability assessment is done and the needs of various populations including pastoralist groups are taken into account.

However, when displacement occurs, it is important that humanitarian assistance is adequately provided. Such assistance must be tailored to the needs of those displaced and should cater for various populations. Moreover, it is essential that such assistance is not merely conceived as a stop gap measure, but is geared towards advancing self-reliance for those displaced. Fourth, it is important that persons affected by climate change are properly documented. Documentation is essential to ensure adequate planning for the needs of those displaced within this context. The Kampala Convention emphasises documentation for IDPs requiring states to 'create and maintain an up-dated register of all internally displaced persons within their jurisdiction or effective control.'[66] Having such documentation process will enhance adequate planning for IDPs in the provision of post-disaster reconstruction and rehabilitation.

Moreover, there must be adequate remediation. The Kampala Convention emphasises the need for states to 'make reparation' for IDPs for 'damage when such a State Party refrains from protecting and assisting internally displaced persons in the event of natural disasters.'[67]

B. Technology-Induced Displaced Persons

In relation to technology-induced displacement, it is essential that a starting point is the furtherance of responsible use of technology as a prevention mechanism. What such responsible use should be geared towards in the context of technology-enabled

---

[65] *Kampala Convention*, art 4(2).

[66] *Kampala Convention*, art 13(1).

[67] *Kampala Convention*, art 12(3).

generalized violence is online safety. States must ensure that such online safety measures are enforced through laws, for instance, that counter cyberbullying and electronic aggression. It is pertinent that these measures are also reflected in Internet Service Provider (ISP) regulations. Further, it is crucial that adequate institutional responses are developed to implement these measures including the creation of units within national ICT structures or specialized cells to combat these forms of violence. In situations of technology-enabled airstrikes, responsible use must be advanced through enhanced target detection which are adequately fitted in Unmanned Aerial Vehicles (UAVs). Such enhanced detection may be enabled through machine learning processes such as neural networks. It is important that UAVs are adequately fitted with such facility to avoid civilian casualties including population displacement. Moreover, these facilities must be vigorously tested and subjected to human control.

There is also the need for regular monitoring and strong reporting systems. Structures must be established to ensure that this is be done in safety and in the furtherance of the right to security of persons. Particularly in situations of technology-enabled generalized violence, it is imperative that states establish specific units within existing structures equipped with competent hotline program analysts who can investigate and analyse reports and threats. For technology-enabled airstrikes, regular monitoring and strong reporting systems must also be ensured. For instance, regular monitoring of the UAVs must be ensured and there must be adequate reporting systems that document the extent of impact of these UAVs in order to ensure enhancements to such systems and avoid situations where civilian casualties occur.

However, where internal displacement occurs, it is imperative that those displaced are provided with prompt specialized medical attention and also granted adequate psychosocial support. This is crucial to ensure that the impact of such situations on them are adequately accessed and addressed. There must also be adequate rehabilitation prioritising safety and security. Humanitarian assistance must be provided and in situations where states do not have adequate capacities for responding to such situations, it is crucial that collaborative efforts are pursued with relevant specialised organisations at both local and international levels.

C. Xenophobia-Induced Displaced Persons

From the start, it is useful to reflect on whether persons displaced due to xenophobia may be regarded as IDPs. Evidently, this question resonates from the fact that IDPs are sometimes regarded as citizens[68] and this even features in some legislations.[69] However, the description of an IDP as 'persons or groups of persons who have been

---

[68] Macaya (2018).
[69] Sudan: The National Policy for Internally Displaced Persons (IDPs) (2009).

forced or obliged to flee their homes or places of habitual residence'[70] points to the fact that nationality or citizenship are not preconditions for protection and assistance.[71] What is imperative is that there has been 'involuntary or forced movement, evacuation or relocation of persons or groups of persons within internationally recognized state borders'.[72]

Given that this form of internal displacement is not permissible, an essential starting point is prevention. In preventing xenophobia-induced displacement, social cohesion measures should be established across various communities within the state. The establishment of this measures may be outlined in normative frameworks including national action plans. Social cohesion measures should be widely prevalent both in societies where xenophobic narratives are in their latent stages and in regions where these narratives are pervasive.

In particularly volatile regions, it is important that such measures are accompanied by strong institutional measures to ensure that these narratives are curtailed and that accountability is advanced. It is also crucial to ensure that early warning systems are established to adequately monitor regions where trends are emerging in order to provide prompt response. Moreover, states must establish strong accountability measures to ensure that actors, public or private that engage in such rhetoric are brought to account. This may be through laws that criminalizes the incitement of xenophobia and includes within this context generalized violence. Moreover, discriminatory policies against specific groups of foreign nationals must be avoided and specifically, this prohibition must be reinforced in legislation at various levels of governance within societies.

Where displacement occurs, it is pertinent that there are adequate structures in place to provide prompt response to those displaced. Access to these structures may be facilitated through the establishment of hotlines for the provision of rapid humanitarian assistance. Moreover, it is imperative that those displaced are also assisted in getting access to specific services, when requested. For instance, where those displaced request access to the consulate of their countries, states must ensure that such access is facilitated. And states must make adequate provisions that ensure that those displaced are ensured protection. Their rights to seek safety in another part of the state or leave the country must be protected. This is reinforced both in the Guiding Principles and the Kampala Convention. However, states should pursue long-term social cohesion measures to ensure that such xenophobic rhetoric are prevented and to safeguard those displaced from future displacement. As such, long-term solutions to displacement should be advanced to ensure safety and advance the well-being of IDPs upon return, resettlement or reintegration. It is also important that they are compensated for losses incurred through adequate remediation processes.

---

[70] *Kampala Convention*, art 1(k).

[71] For further discussion on the legal description of an IDP, see Adeola (2020, 6–7).

[72] *Kampala Convention*, art 1(l).

D. Harmful Practices-Induced Displaced Persons

Harmful practices are explicitly prohibited under the Kampala Convention, hence a pertinent starting point for understanding the content of protection for IDPs in this context is the prohibition of the root cause. Such prohibition must resonate in national legal provisions and must be enforced by competent and duly capacitated institutions within states. The existence of these laws must be accompanied by a sustained advocacy on the provisions to ensure adequate understanding and to prevent such forms of practices. In this context, leveraging on alternative institutions such as traditional and faith-based institutions and civil society at large is crucial, not least, given the nature of this practice and the contexts from which it may emerge.

Moreover, there is a need to ensure that community-based monitors are also placed within societies to ensure that such practices are not perpetrated. Such community-based monitors will imply leveraging on existing societal structures and creating a bottom-up solution with community members as protection administrators. Having such community-based monitors will serve to ensure that community members are empowered to prevent the spread of such practice and displacement that may occur due to its occurrence. Moreover, having community-based monitors will serve to ensure that there is an effective complement to formal state structures towards preventing such practices and consequent displacement that may emerge from these contexts.

However, where such situations of internal displacement occur, affected persons must be provided with adequate psychosocial rehabilitation. This must take into account the nature of the practice and attendant consequences on the affected persons. Such support should not be conceived solely as a short-term measure but rather, it must be provided to the extent to which it is required by those displaced. Hence, it is imperative that measures be put in place to sustain such systems of support for affected persons. Moreover, there is a need to also provide adequate humanitarian assistance and ensure that durable solutions are geared towards securing the safety of these persons in places of resettlement, return or reintegration. Overall, there is a need for adequate follow-up mechanisms to ensure that these persons are not further affected and situations of internal displacement do not occur following the initiation of long-term solutions.

E. Generalized Violence-Induced Displaced Persons

There are specific due process requirements in situations of generalized violence. Given that these are non-permissible forms of internal displacement, prevention is the fundamental starting point of advancing protection for IDPs. As such, the first due process requirement is the development of conflict early warning systems to ensure that such violence is duly addressed. The Guiding Principles emphasises the pertinence of this in providing that IDPs shall be protected from 'acts of violence intended to spread terror'.[73] Preventing acts of terror will require that such systems are duly financed and resourced with adequate capacities. Having such early warning systems can aid the monitoring of latent zones to ensure that populations are adequately secured from situations of internal displacement.

As a matter of prevention, there is also a need for states to tackle the systemic causes of such violence. In the context of gang violence, for instance, addressing systemic issues may require a range of measures to address the underlying root cause of such violence. Depending on the nature of the cause and the society in which such forms of violence exists, measures may range from addressing unemployment concerns to combatting prevalent socio-economic disparities. Moreover, there may be a need to address pervasive insecurity where such violence is a consequence of the absence of institutional safety measures. In the context of electoral violence, it may be crucial to ensure that political party rhetoric that may incite such violence are prohibited and electoral candidates are mandated to foster social cohesion. In the context of violent extremism, measures may range from tackling marginalization to widespread sensitization against extremist views.

But where displacement occurs, displaced persons must be properly rehabilitated. Such rehabilitation must be done in safety and with adequate security measures in place to prevent further incidences. Psychosocial support must be prioritised. Humanitarian assistance must also be given to the affected persons and they must be involved in planning durable solutions. Moreover, adequate remediation should be carried out to ensure that they are compensated for their losses and provided sufficient reparations. It may also be the case that transitional justice processes are initiated following situation of internal displacement, in such case, IDPs must be properly incorporated into such processes. Moreover, states must ensure that the provisions of the IDP laws are incorporated into peace processes in ensuring durable solutions.[74]

F. Development Projects-Induced Displaced Persons

Prior to the execution of a development project, there are certain due process requirements. First, feasible alternatives should be considered together with persons likely to be displaced. There are six alternatives that may be considered in this regard.

First, is the option of not implementing the project at all. But where this is not feasible given the imperative necessitating the execution of the development project, another feasible alternative that may be considered is the possibility of initiating the project in a different location. This is relevant in order to avoid displacement altogether. But where this is not feasible, the project may be done collaboratively

---

[73] *Guiding Principles*, principle 11(2)(c).

[74] Article 3(2)(e) of the Kampala Convention provides that states shall '[e]ndeavour to incorporate the relevant principles contained in this Convention into peace negotiations and agreements for the purpose of finding sustainable solutions to the problem of internal displacement.' *Kampala Convention*, art 3(2)(e).

with persons likely to be displaced. This alternative is particularly relevant to situations where projects are to be carried out on indigenous territories. But where this is not also feasible, there is the alternative of adopting strategies to minimise displacement such as, for instance, altering the project design. However, where this is not feasible, the possibility of entering into a lease agreement with persons likely to be displaced may be considered so as to ensure that they retain ownership over their properties and territories. However, where these alternatives are not feasible, then voluntary compensation should be pursued. It is pertinent that persons displaced within this context decide on their compensation. And as such, there must not be a superimposition on their will to decide. The consideration of these feasible alternatives must be with the full information and consultation of these persons. Moreover, prior impact assessment of the socio-economic and environmental impacts of these projects must be examined.

4.3 Conclusion

Globally, the response to internal displacement has led to the emergence of normative standards. At the UN level, this is reflected in the Guiding Principles. Within the African context, this emanates from the provision of the Kampala Convention.

Both norms reinforce the prevention of arbitrary displacement and the right of persons to be protected from arbitrary displacement. This chapter argues that there are two imperatives of this protection: displacement must be permitted in law and due process must be observed. There are three pertinent dimensions of due process: general guarantees, group-based guarantees and cause-specific guarantees. While general guarantees resonate from wide-ranging provisions in these frameworks, group-based guarantees relate to guarantees from norms that protect specific populations. However, cause-specific guarantees are guarantees that resonate from the specific root cause of internal displacement.

Within the context of climate change, there are certain imperatives: adequate planning ahead of climate-related disasters, engagement with local communities, humanitarian assistance, documentation of those displaced and adequate remediations. With respect to technology, it is pertinent that there is responsible use as a preventive mechanism. Further, there needs to be regular monitoring and robust reporting systems. However, where displacement occurs, specialized medical attention and adequate psychosocial support must be provided. In situations of xenophobic, social cohesion measures must be put in place and accompanied by strong institutional measures. Moreover, there is a need for strong accountability measures including law and institutions. Displaced persons must also have access to specific services. For non-nationals, for instance, this may include facilitating access to the consulate of their countries. Further, durable solutions must be fostered with long-term social cohesion. With respect to harmful practices, it is crucial that harmful practices are prohibited through normative frameworks coupled with sustained advocacy. Where internal displacement occurs, there needs to be adequate psychosocial support and humanitarian assistance. Further, durable solutions must be geared towards ensuring the safety of displaced persons. Moreover, there is a need for adequate follow-up mechanisms. In situation of generalized violence, states should develop conflict early warning systems as a preventive measure. Further, pervasive insecurity must be addressed. Where displacement occurs, those displaced must be properly rehabilitated in safety and security. Psychosocial support must also be ensured and humanitarian assistance provided. Moreover, there is a need for adequate remediation.

Within the context of development projects, this requires considering feasible alternatives with persons likely to be displaced, ensuring that they are fully informed and duly consulted. Moreover, project socio-economic and environmental impacts need to be assessed.

References

Abebe AM (2016) The emerging law on forced migration in Africa: development and implementation of the Kampala convention on internal displacement. Routledge, London

Adeola R (2019) The impact of the African Union convention on the protection and assistance of internally displaced persons in Africa. Afr Hum Rights Law J:591–607

Adeola R (2020) The internally displaced person in international law. Edward Elgar, Cheltenham

African Charter on Human and Peoples' Rights, adopted by the Organisation of African Unity, OAU Doc CAB/LEG/67/3 rev 5 (27 June 1981)

African Charter on the Rights and Welfare of the Child, adopted by the Organisation of African Unity, OAU Doc CAB/LEG/153/Rev. 2 (11 July 1990)

African Union Convention for the Protection and Assistance of Internally Displaced Persons in Africa, adopted on 23 October 2009

Anghie A (2004) Imperialism, sovereignty and the making of international law. Cambridge University Press, Cambridge

AU Executive Council, Decision on the situation of refugees, returnees and displaced persons, AU Doc EX/CL/127 (V)

AU Executive Council, Decision on the meeting of experts on the review of OAU/AU treaties, AU Doc EX.CL/Dec. 129 (V)

AU Protocol on Amendments to the Protocol on the Statute of the African Court of Justice and Human Rights (2014)

Bellamy AJ (2015) The responsibility to protect: a defense. Oxford University Press, Oxford

Bernstein J, Bueno O (2007) The Great Lakes Process: new opportunities for protection. Forced Migrat Rev 29:73

Beyani C (2006) Recent developments: the elaboration of a legal framework for the protection of internally displaced persons in Africa. J Afr Law 50(2):187–197

Beyani C (2007) Pact on security, stability and development in the Great Lakes region. Int Leg Mater 46(2):173–184

Breau S (2016) The responsibility to protect and international law: an emerging paradigm shift. Routledge, London

Cantor DJ (2018) Returns of internally displaced persons during armed conflict: international law and its application in Columbia. Brill-Nijhoff, Leiden

Cardona-Fox G (2019) Exile within borders: a global look at commitment to the international regime to protect internally displaced persons. Brill, Leiden

Cohen R, Deng F (1998) Masses in flight: the global crisis of internal displacement. Brooking Institution Press, Washington, DC

Convention on the Elimination of All Forms of Discrimination Against Women (1979)

Convention on the Rights of the Child, adopted by the UN General Assembly Resolution 44/25, UN Doc A/RES/44/25 (20 November 1989)

Declaration on Peace, Security, Democracy and Development (2004)

Democratic Republic of Congo: Bill providing protection and assistance to internally displaced persons (2014)

Deng FM (2001) The global challenge of internal displacement. Wash Univ J Law Policy 5(1):141

Elettronica Sicula S.p.A. (ELSI) (United States of America v. Italy) (Judgement) [1989] ICJ Rep 15

Evans G (2015) The evolution of the responsibility to protect: from concept and principle to actionable norm. In: Thakur R, Maley W (eds) Theorising the responsibility to protect. Cambridge University Press, Cambridge

Genser J, Cotler I (eds) (2012) The responsibility to protect: the promise of stopping mass atrocities in our time. Oxford University Press, Oxford

Groth L (2011) Engendering protection: an analysis of the 2009 Kampala Convention and its provision for internally displaced women. Int J Refugee Law 23(2):1

Guzman AT, Meyer TL (2010) International soft law. J Legal Anal 2(1):171–225

Inter-American Convention on Protecting the Human Rights of Older Persons (2015)

Kälin W (2001) 'How hard is soft law? The guiding principles on internal displacement and the need for a normative framework' Presentation at roundtable meeting, Ralph Bunche Institute for International Studies, CUNY graduate Centre, 19 December

Kälin W (2014a) The guiding principles on internal displacement and the search for a universal framework of protection for internally displaced persons. In: Chetail V, Bauloz C (eds) Research handbook on international law and migration. Edward Elgar, Cheltenham

Kälin W (2014b) Internal displacement. In: Fiddian-Qasmiyeh E, Loescher G, Long K, Sigona N (eds) The Oxford handbook of refugee and forced migration studies. Oxford University Press, Oxford

Kleine-Ahlbrandt ST (2004) The protection gap in the international protection of internally displaced persons: the case of Rwanda. Graduate Institute of International Studies Working Papers

Kuwali D, Viljoen F (eds) (2014) Africa and the responsibility to protect: article 4(h) of the African Union Constitutive Act. Routledge, London

Macaya J (2018) IDPs: invisible citizens or blinded authorities? The New Context Aug 29

Mutua MW (2000) What is TWAIL? American Society of International Law, Proceedings of the 94th Annual Meeting 31

Niger: Law No. 2018-74 of 10 December 2018, relating to the Protection and Assistance of Internally Displaced Persons (2018)

Nigeria: Rights of Internally Displaced Persons (IDPs) Bill (2016)

OAU Convention Governing the Specific Aspects of Refugee Problems in Africa (1969)

Okafor OO (2008) Critical third world approaches to international law (TWAIL): theory, methodology, or both? Int Commun Law Rev 10:371–378

Optional Protocol to the Convention on the Rights of the Child on the Involvement of Children in Armed Conflict, adopted by UN General Assembly Resolution 54/263, UN Doc A/RES/54/263 (25 May 2000)

Orchard P (2019) Protecting the internally displaced: rhetoric and reality. Routledge, London

Oslo declaration and plan of action on the plight of refugees, returnees and displaced persons in southern Africa, adopted at the International conference on the plight of refugees, returnees and displaced persons in southern Africa in Norway (22–24 August 1988)

Pact on Security, Stability and Development in the Great Lakes Region (2006)

Phuong C (2004) The international protection of internally displaced persons. Cambridge University Press, Cambridge

Prevention, Protection and Assistance to Internally Displaced Persons and Affected Communities Act (2012)

Protocol Additional to the Geneva Conventions of 12 August 1949 and Relating to the Protection of Victims of Non-international Armed Conflict (8 June 1977)

Protocol on Free Movement of Persons, Right of residence and right of establishment in Africa, adopted by the AU heads of states and government thirtieth ordinary session in Addis Ababa, Ethiopia (January 2018)

Protocol to the African Charter on Human and Peoples' Rights on the Rights of Older Persons in Africa (2016)

Protocol to the African Charter on Human and Peoples' Rights on the Rights of Persons with Disabilities in Africa (2018)

Rome Statute of the International Criminal Court (1998)

Somalia: National Policy on Refugee-Returnees and Internally Displaced Persons (2019)

South Sudan: Protection and Assistance to Internally Displaced Persons Act 2019 (Draft)

Sudan: The National Policy for Internally Displaced Persons (IDPs) (2009)

Treaty Establishing the African Economic Community, adopted by the Heads of State and Government of Member States of the Organisation of African Unity in Abuja, Nigeria, 3 June 1981

UN Commission on Human Rights, Addendum (11 February 1998) Guiding principles on internal displacement. Report of the representative of the Secretary-General, Mr. Francis M. Deng, submitted pursuant to Commission on Human Rights resolution 1997/39, UN Doc. E/CN.4/1998/53/Add.2

UN Commission on Human Rights internally displaced persons, UN Doc E/CN.4/RES/1992/73 (5 March 1992)

UN convention on the rights of persons with disabilities (2006)

UN declaration on the rights of indigenous peoples (2007)

UN General Assembly. International conference on the plight of refugees, returnees and displaced persons in Southern Africa: Report of the UN Secretary-General, UN Doc A/43/717 (19 Oct 1988a)

UN General Assembly. International conference on the plight of refugees, returnees and displaced persons in southern Africa, UN Doc A/RES/43/116 (8 Dec 1988b)

United Nations principles for older persons (1991)

World Summit Outcome, adopted by the UN General Assembly, Resolution 60/1, UN Doc A/RES/60/1 (24 October 2005)

Zambia: Guidelines for the compensation and resettlement of internally displaced persons (2013)

Chapter 5
Conclusion and Recommendations

5.1 Conclusion

The issue of internal displacement is a pertinent challenge globally. While significant attention has been given to conflict as the principal driver of internal displacement, not much has emerged with respect to other root causes of internal displacement
that have become emerging challenges. This book examines six principal drivers:
climate change, technology, xenophobia, harmful practice, generalized violence
and development project.

The nexus between climate change and internal displacement is undoubtedly an
imperative that has emerged and is increasingly gaining relevance in discussions on
addressing the multifaceted impact of climate change. This book identifies four
dimensions: sudden on-set disaster, slow on-set disasters, climate-related conflict
and climate development-induced displacement. The nexus between technology
and internal displacement has also become pertinent given the global drive towards
technology advancement and the increasing impact of technology as an enabler of
development in various societies. This chapter identifies two principal dimensions
of the nexus between technology and internal displacement: technology-enabled
attacks and technology-enabled generalized violence. Xenophobia has become a
pertinent root cause of internal displacement given the emergent rhetoric against
migrants in various countries and the legitimation of populism across various societies. Africa is no exception. This book spotlights two principal dimensions of this
nexus: xenophobia-induced generalized violence and implementation of discriminatory policies against specific groups of foreign nationals.

The prevalence of harmful practices in various parts of the continent makes paying attention to this issue an imperative in providing solution to the various ramifications of internal displacement. This book examines two pertinent dimensions of
this: gender-based and group-based practices. Generalized violence and internal
displacement have also become a pertinent challenge in various parts of the

© The Author(s), under exclusive license to Springer Nature
Switzerland AG 2021
R. Adeola, *Emerging Issues in Internal Displacement in Africa*, SpringerBriefs
in Law, https://doi.org/10.1007/978-3-030-64562-5_5

continent given the prevalence of this form of violence and its dimensions in many parts of the continent. This book examines five pertinent dimensions of this: gang violence, electoral violence, interethnic clashes, violent extremism and technology-enabled generalized violence. Development projects have also become a prevalent root cause of internal displacement given the prevalence of these projects in various parts of the continent since the early period of independence in various parts of the continent. This book considers six dimensions of this nexus: dams, natural resource extraction, agricultural investment, urban renewal and climate-related projects.

In response to the issue of internal displacement, the *UN Guiding on Internal Displacement* (Guiding Principles) has emerged as a global soft law standard.[1] However, this soft norm has significantly led to the development of other norms at regional and national levels over the last three decade, including the *Protection and Assistance of Internally Displaced Persons in the Great Lakes Region*.[2] Within the African context, a pertinent manifestation of this is through the formation of the African Union Convention for the Protection and Assistance of Internally Displaced Persons in Africa (Kampala Convention).[3] Both instruments expansively describe IDPs, providing for a non-exhaustive list of causes. As such, recognising that internal displacement may arise from various root causes besides those listed in the description, i.e. 'armed conflict, situations of generalized violence, violations of human rights, natural or human-made disasters'.[4] The Kampala Convention reinforces the provision of the Guiding Principles on protection of populations from arbitrary displacement. This provision is an integral cornerstone of the law on internal displacement. There are two dimensions of this obligations: the root cause of internal displacement must be permissible under international law. Second, due process must be followed.

Considering the various root causes of internal displacement, a pertinent question that emerges in relation to the first dimension of the obligation is whether these root causes are permissible. Under international law, the permissibility of displacement resonates in two of these emerging contexts: climate change and development projects. In relation to climate change, the Kampala Convention links this form of displacement to natural disasters. Internal displacement within the context of natural disasters is permissible where it is for the safety and health of the populations.[5] Internal displacement in the context of development is also permissible, although the Kampala Convention obligates states to prevent this form of displacement: 'as much as possible'.[6] However, internal displacement premised on harmful practices and generalized violence are explicitly prohibited. Although the Kampala Convention does not explicitly mention technology and xenophobia, these root

---

[1] UN Commission on Human Rights, Addendum (1998).

[2] *Protocol on the protection and assistance of internally displaced persons* (2006).

[3] *African Union convention for the protection and assistance of internally displaced persons in Africa* (2009).

[4] *Guiding Principles*, para 2; *Kampala Convention*, art 1(k).

[5] *Kampala Convention*, art 4(4)(f).

[6] *Kampala Convention*, art 10.

causes are closely linked to other non-permissible forms such as violations of human rights and generalized violence. And the Kampala Convention prohibits internal displacement where it is premised on 'any act, event, factor, or phenomenon of comparable gravity'.[7]

However, where internal displacement occurs, there is a responsibility to ensure that due process is followed. This book reinforces three due process requirements: general guarantees, cause-specific and group-based requirements. General guarantees relate to wide-ranging provisions under the Guiding Principles and the Kampala Convention including the protection of IDP rights and the provision of durable solutions. Cause-specific guarantees relate to general requirements that are imperative in the furtherance of protection within emerging contexts.

In situations of climate change, there is a need for adequate planning ahead of climate-related disasters through the development of early warning systems, engagement with local communities, humanitarian assistance, documentation of IDPs and adequate remediation. In situations of technology, an essential guarantee is ensuring the responsible use of technology as a prevention mechanism. In the context of technology-enabled generalized violence this will require online safety. In situations of technology-enabled airstrikes, responsible use will require enhanced target detection that are adequately fitted in unmanned aerial vehicles (UAVs). There is also a need for regular monitoring and strong reporting systems. When displacement occurs, specialized medical attention must be given along with adequate psychosocial support. Moreover, humanitarian assistance must be provided with the support of specialised organisations where a state does not have the adequate capacity to provide it.

In situations of xenophobia, it is pertinent that social cohesion measures be pursued as a preventive measure. These should be accompanied by strong institutional measures in volatile regions, as such, in areas where xenophobic rhetoric is prevalent and where the likelihood of an outbreak of violence is evident. States must also establish strong accountability measures including law and institutional measures. In situations where displacement occurs, humanitarian assistance should be promptly provided. Moreover, displaced populations should have access to specific services. For non-nationals, this may require facilitating access to their consulates, when requested. Moreover, durable solutions should be advanced with long-term social cohesion. With regards to harmful practices, there should be prohibition of such practices through legislations accompanied with sustained advocacy through institutions such as traditional and faith-based institutions. Moreover, community-based monitors must also be established to advance this objective. However, where internal displacement occurs, it is imperative that adequate psychosocial support is provided. Humanitarian assistance must also be provided and durable solutions must be geared towards ensuring the safety of those displaced. Moreover, adequate follow-up mechanisms must be ensured.

In situations of generalized violence, it is crucial that conflict early warning systems are developed to prevent such incidences. Moreover, states should tackle the

---

[7] *Kampala Convention*, art 4(4)(h).

systemic causes of such violence as an integral part of prevention. Addressing pervasive insecurity is also imperative. However, where displacement occurs, proper rehabilitation in safety and security of displaced populations must be ensured. Psychosocial support must be made available, humanitarian assistance must be ensured and adequate remediation must also be provided. In situations of development projects, it is imperative that there is a consideration of feasible alternatives to the development projects with persons likely to be displaced. This must be done with the full information and consultation of these persons. Also, there must be a prior impact assessment of the socio-economic and environmental impacts of the development project.

In addition, there are group-based guarantees that must be ensured. These guarantees will require that the norms relating to the various categories such as women, children, persons with disabilities, older persons and indigenous peoples are observed.

5.2 Recommendations

Translating the regional normative provisions into concrete nation-level actions in addressing these emergent issues require specific measures geared towards protecting IDPs. This section examines seven recommendations.

First, there is a need for data on the emergent issues in internal displacement. Building action towards concrete normative protection requires the existence of credible data on the nexus between the various issues and internal displacement. Data is crucial for intervention at various governance levels. In addressing the data gap, it is important that the voices of the displaced are integrated.

Linked to this is the importance of research on internal displacement, particularly in relation to the specific vulnerabilities of affected populations. The essence of research in this area emanates from the differentiated impacts of these emerging issues in the IDP context on various categories such as children and persons with disabilities. Intersectionalities are also imperative considerations, particularly where vulnerabilities are compounded by the existence of such. For instance, an internally displaced girl may experience compounded vulnerability in internal displacement more severely where adequate measures are not undertaken to ensure sustainable protection that takes into account the existence of gender and age.

Further, there is a need to strengthen compliance mechanisms at the national level to afford IDPs effective protection that can concretely be asserted. The essence of compliance cannot be overemphasised given that through compliance, treaty provisions are achieved. As the primary test of norms is at national level, compliance mechanisms are imperative in translating regional provisions into concrete actions in the furtherance of protection for IDPs. While these mechanisms are imperative at various levels, specific national-level compliance mechanisms that cut across institutional mandates in the areas of peace, migration, development, climate change, innovation, social protection and human rights, are imperative.

Moreover, it is important to address issues that impede on protection and humanitarian assistance such as the absence of rule of law, democratic deficits and governance deficiencies. Evidently, these are issues that affect adequate safeguards for IDPs and even where there is international support – if these issues persist – they will impede on the impact of interventions. Addressing these issues necessitates the establishment of an effective institutional structure for accountability and sustained support of an active civic space.

Also, it is important that IDP protection is achieved through interventions that cut across the humanitarian, development and governance dimensions, with the later focusing on the underlying causes for some of these emerging issues such as generalized violence and xenophobia and also focusing on deficiencies in existing national structures that impede on adequate protection and assistance to IDPs.

In addition, there is a need to build evidence-based good practices that are effective and relevant. Good practices are important given the fact that they provide replicable suggestions. Such practices may emerge from community-based adaptation strategies, normative formations at sub-national level, resilience measures, mechanisms for the integration of specific groups and participatory processes. Moreover, good practices may also emerge from positive traditional values that are integrated into law and policy formations geared towards the protection of internally displaced persons. Good practices are important to provide a range of possibilities that may be adopted to translate norms into national actions.

Overall, states should develop national action plans on internal displacement, drawing on these normative provisions and taking into account the various dimensions of internal displacement. National action plans on these issues are imperative to ensure the prioritisation of the emergent forms and to advance guidance to various stakeholders in relevant sectors at various levels of governance in order to promote a whole-of-society approach. Such plans, over a given period of time, will ensure that institutional machineries at the national level have a common mission and that parallel actions are not proliferated.

5.3   Implication for Refugees and Migration Studies

There are some implications of these emerging contexts for the field of refugee and migration studies. International refugee law is quite specific in its definition on refugees while the law on IDPs is quite expansive in its description.[8] Incorporating these

---

[8] Article 1 of the UN Refugee Convention Relating to the Status of Refugees recognises refugee as 'well-founded fear of being persecuted for reasons of race, religion, nationality, membership of a particular social group or political opinion, is outside the country of his nationality and is unable or, owing to such fear, is unwilling to avail himself of the protection of that country; or who, not having a nationality and being outside the country of his former habitual residence as a result of such events, is unable or, owing to such fear, is unwilling to return to it.' See UN *refugee convention relating to the status of refugees* (1951), art 1.

emerging factors in refugee law raises conceptual challenges. The kind of challenge this poses for international refugee law has emerged, for instance, within the context of climate change. In international refugee law, the notion of international protection for persons displaced across borders by climate change, has been significantly debated.

However, at the African regional level, there are emerging discussions on the pertinence of legal protection for 'climate refugees', with an expressive indication that the 'refugee definition in the 1969 OAU Refugee Convention may allow decision-makers to recognize refugee status in the context of climate change.'[9] However, what might be regarded as 'events seriously disturbing public order' and whether this will involve situations of natural disasters, including climate change has not been conclusive. Whether the notion of climate change will find expression in the regional refugee legal framework remains to be seen. However, in migration studies, the nexus between climate change and migration has received considerable attention and gained notable traction both in global and regional processes. In the Global Compact on Safe, Orderly and Regular Migration, for instance, this nexus resonates.[10] In the AU Migration Policy Framework for Africa and Plan of Action, the link between climate change and migration is also reinforced.[11]

With regards to technology, much of the discussion in refugee and migration studies have been on the positive role of technology in protecting persons within these contexts and on the negative impact of technology mostly in regards to profiling and fuelling anti-foreigner sentiments. But there is yet to emerge significant discussion on technology as a driver of cross-border movements that may fall within the refugee and migration studies. Similarly, a significant body of research has emerged on xenophobia against migrants and refugees and how this results in marginalization and significant exclusions, however, not much research exists on xenophobia as a driver of migration and mobility. An extensive body of work is beginning to examine the linkage between harmful practices and refugee law, mostly with respect to issues of female genital mutilation.[12] Similarly, within the Latin American context, scholars have examined generalized violence as a root cause of refugee movement.[13] But research on development projects within the context of refugee and migration studies has been scarce, perhaps given that much of these movements remain within state borders. However, there may be occasions, for instance, where such situations may drive displacement across borders, for instance, where these projects are done along border regions.

---

[9] During a Roundtable on addressing root causes and achieving durable solutions in Africa, organized by the African Union Commission in 2019 as part of a series of evident on the AU's theme of the Year on Refugees, Returnees and IDPs, this observation was made. See African Union *Summary conclusions: Roundtable on addressing root causes and achieving durable solutions in Africa* (2019).

[10] *Global compact on safe, orderly and regular migration* (2018).

[11] *AU migration policy framework for Africa and plan of action* (2018–2030).

[12] See United Nations High Commissioner for Refugees 2009.

[13] Zolberg et al. (1989, 269).

Interrogating these issues in refugee and migration studies is imperative in developing knowledge and guidance on adequate legal response beyond the dimensions of forced movements within state borders.

References

African Union (2019) Summary conclusions: roundtable on addressing root causes and achieving durable solutions in Africa. Addis Ababa, Ethiopia, 9 Feb 2019

African Union convention for the protection and assistance of internally displaced persons in Africa, adopted on 23 October 2009

African Union migration policy framework for Africa and plan of action (2018–2030)

Global compact on safe, orderly and regular migration, adopted by the intergovernmental conference on the global compact for migration in Marrakesh, Morocco, 10–11 Dec 2018

Protocol on the protection and assistance of internally displaced persons (2006)

UN Commission on Human Rights, Addendum (1998) Guiding Principles on Internal Displacement. Report of the Representative of the Secretary-General, Mr. Francis M. Deng, submitted pursuant to Commission on Human Rights resolution 1997/39, UN Doc. E/CN.4/1998/53/Add.2, 11 Feb 1998

UN High Commissioner for Refugees (2009) Guidance note on refugee claims relating to female genital mutilation

UN Refugee convention relating to the status of refugees (1951)

Zolberg AR, Suhkre A, Aguayo S (1989) Escape from violence: conflict and the refugee crisis in the developing development. Oxford University Press, Oxford

Index

A
Adaptation, 21
Affluent societies, 9
Africa
  decolonisation, 3
  OAU, 3
African Union (AU) Convention, 2, 44–47, 64
Agrarian revolution, 8, 9
Agricultural investment, 31, 34, 64
Aircrafts, 23
Anthropocene Era, 6
Anti-foreigner sentiments, 68
Anti-imperialism, 3
Anti-migrant sentiments, 24
Arbitrary, 52
Armed conflicts, 12, 13, 43, 64
Artificial intelligence, 9
Attitudinal prejudices, 25
AU Executive Council, 46
AU Migration Policy Framework, 68
Automation, 9

B
Big data, 9
Breast ironing, 27, 28
Brookings-LSE Project, 2
Brookings-SAIS Project, 2

C
Cartagena Declaration, 12
Cataclysms, 13
Cause-specific guarantees, 53
Child marriage, 28

Climate change, 63, 65
  climate development-induced
    displacement, 21
  climate-related conflicts, 21
  cyclical and seasonal movements, 21
  DRC, 22
  droughts conditions, 21
  environmental refugees, 8
  global surface temperature, 5
  global temperatures, 19
  global validation, 7
  hothouse state, 6
  IPCC, 6, 7
  IPPF, 22
  minimalists/sceptics, 7
  NFC, 22
  NRMP, 21
  Paris Agreement, 19
  SIDS, 6
  slow on-set disasters, 20
  sudden on-set disasters, 20
  UNEP, 19
Climate change-induced displaced persons, 54
Climate development-induced displacement,
    21, 32, 63
Climate displaced persons, 8
Climate-induced stresses, 21
Climate migrants, 8
Climate-mobility nexus, 8
Climate-related conflicts, 21, 32, 63
Climate-related disasters, 65
Climate refugees, 8
Climate variability, 7
Coercive social controls, 12
Cold War, 3

Committee on the Elimination of Discrimination against Women (CEDAW), 11
Compelling and overriding public interests, 53
Cultural facet, 3
Cultural traditions, 11
Cyclical and seasonal movements, 21

D
Dams, 31, 34, 64
Decolonisation, 3
Delegation theory, 49
Democratic Republic of Congo (DRC), 22
Desertification, 20
Development-induced displacement, 13
Development projects, 60, 63, 64, 66, 68
   agricultural investment, 31
   dams, 31
   economic growth, 13
   global estimate, 31
   natural resource extraction, 31
   principal, 13
   and rights, 13
   societal order, 13
   urban renewal and climate-related projects, 32
Development projects-induced displaced persons, 58–59
Digital age, 32
Digital technology, 10
Drones, 23
Droughts conditions, 21
Due process, 53
Durable solutions, 51, 52

E
Economic facet, 3
Economic growth, 13
Electoral violence, 29, 30, 33, 64
Electrification, 9
Elettronica Sicula S.p.A. (ELSI) (United States of America v Italy), 52
Emerged and normative guidance, 1
Emerging contexts, 2, 33, 41–60, 64, 65, 67
Environmental emigrants, 8
Environmental migrants, 8
Environmental refugees, 8
Etymology, 10
Evidence-based knowledge, 32

F
Female genital mutilation (FGM), 28
Foreign Policy, 23
Forest degradation, 20

Four Geneva Conventions, 45
Free Movement Protocol, 45

G
Gang-related violence, 12
Gang violence, 29, 33, 64
Gender-based practices, 27, 33
General guarantees, 53
Generalized violence, 13, 32, 33, 43, 50, 53, 55, 56, 60, 63–65, 67, 68
   armed conflicts, 12
   electoral violence, 29, 30
   gang violence, 29
   interethnic clashes, 30
   technology-enabled generalized violence, 30
   UN High Commissioner for Refugees' Guidelines on International Protection, 12
   violent extremism, 30
Generalized violence-induced displaced persons, 57, 58
Global Compact on Safe, 7, 68
Global governance, 6
Global rhetoric, 10
Global surface temperature, 5
Global temperatures, 19
Global validation, 7
Great Lakes Protocol, 48, 49, 51, 52
Great Lakes Region, 47, 48
Greenhouse gases, 6
Group-based practices, 28, 29, 33
Group-specific guarantees, 53

H
Habitual residence, 42
Harmful practices, 32, 33, 50, 53, 59, 63–65, 68
   attitudinal orientations, 11
   behavioural patterns, 11
   breast ironing, 27, 28
   child marriage, 28
   cultural traditions, 11
   FGM, 28
   gender-based practices, 27
   group-based practices, 28, 29
   institutional systems, 12
   mannerisms, 11
   regular law, 12
   sexual servitude, 27
   societal norms, 11
   violence, 11
Harmful practices-induced displaced persons, 57

Hothouse state, 6
Human development, 9
Humanitarian assistance, 47, 51, 53, 65, 66
Humanitarian law, 50
Human rights, 24, 43, 52
Humanity, 51

I
Impartiality, 51
Indigenous Peoples' Planning Framework
    (IPPF), 22
Industrial revolution, 9, 10
Interethnic clashes, 21, 30, 33, 64
Intergovernmental Panel on Climate Change
    (IPCC), 6, 7
Internal displacement
    Brookings-LSE Project, 2
    climate change (*see* Climate change)
    emerging root causes, 1, 2
    imperative of law, 2
    principal, 1
Internally displaced persons (IDPs)
    AU Convention, 44–47
    global structures, 2
    ICGLR, 47, 48
    legal protection, 2
    migration studies, 67–69
    protection, 2, 42
    recommendations, 66, 67
    refugees, 67–69
    regional structures, 2
    UN Guiding Principles, 42–44
International common law (ICL) theory, 49
International Conference of the Great Lakes
    Region (ICGLR) IDPs, 47, 48
International Conference on the Plight of
    Refugees, Returnees and Displaced
    Persons in Southern Africa
    (SARRED), 42, 43
International Court of Justice, 52
International human rights, 50
International humanitarian law, 43, 52
International law, 45, 52, 53
International Law Commission (ILC), 3
International refugee law, 67–69
Internet, 9
Internet Service Provider (ISP) regulations, 55
Inter-regionalism, 3
Iron ploughs, 9
Iron-working technology, 9

K
Kakuma refugee camp, 26
Kampala Convention, 2, 3, 64, 65

L
Legal protection
    climate change-induced displaced
        persons, 54
    definition, 3
    development projects-induced displaced
        persons, 58–59
    due process, 53
    durable solutions, 51, 52
    generalized violence-induced displaced
        persons, 57, 58
    harmful practices-induced displaced
        persons, 57
    humanitarian assistance, 51
    ICL theory, 49
    IDPs, 2 (*see also* Internally displaced
        persons (IDPs))
    Kampala Convention, 49
    prevention, 49, 50
    protection, 50, 51
    soft laws, 49
    technology-induced displaced
        persons, 54, 55
    xenophobia-induced displaced
        persons, 55, 56
Legal scholarship, 45
Loss of biodiversity, 20

M
Mannerisms, 11
Maximalists, 7
Migration, 3, 67–69
Minimalists/sceptics, 7
Mitigation, 21
Mobility, 7
Mono-causal argument, 7
Multi-causality, 7

N
National contexts, 33
Nation-state ideological formations, 10
Natural disasters, 13
Natural resource extraction, 31, 34, 64
Natural Resource Management Programme
    (NRMP), 21
Neutrality, 51
New Forest Company (NFC), 22
Non-state actors, 12
Normative frameworks, 45

O
OAU 1969 Refugee Convention, 44
OAU Refugee Convention, 68

Ocean acidification, 20
Online platforms, 11
Orderly and Regular Migration, 7, 68
Organisation of African Unity (OAU), 3, 46

P
Pan-Africanism, 46
Paris Agreement, 19, 32
Policies, 11
Political facet, 3
Potential low-observability, 23
Project-affected persons, 31
Project-affected populations, 13
Psychosocial support, 66

R
Recommendations, 66, 67
Refugee law, 43
Refugees, 67–69
Regionalism, 3

S
Sea level rise, 20
Security operations, 26
Seed drills, 9
Sexual servitude, 27
Sheer gravity, 47
Slash-and-burn cultivation, 8
Slow on-set disasters, 20, 32, 63
Small island developing states (SIDS), 6
Social facet, 3
Social formations, 11
Social media, 11, 24
Social sciences, 13
Societal norms, 11
Societal order, 13
Soft laws, 49
Sudden on-set disasters, 20, 32, 63
Sustainable solutions, 1

T
Technology
  agrarian revolution, 8, 9
  aircrafts, 23
  artificial intelligence, 9
  automation, 9
  big data, 9
  digital, 10
  Foreign Policy, 23

industrial revolution, 9–10
  iron-working, 9
  migration, 23
  mobility, 23
  negative and positive impacts, 10
  social media, 24
  technology-enabled attacks, 23
  technology-enabled generalized
    violence, 24
  UAVs, 23
Technology-enabled airstrikes, 65
Technology-enabled attacks, 23, 33, 53, 63
Technology-enabled generalized violence, 24,
  30, 33, 53, 63–65
Technology-induced displaced persons,
  54, 55
Telecommunications, 9
Terror/indiscriminate attack, 12
Thermodynamics, 9
Traditional and faith-based institutions, 65
Tri-dimensional categorisation, 8

U
UN 1951 Refugee Convention, 43, 44
UN Commission on Human Rights, 42
UN Guiding on Internal Displacement
  (Guiding Principles), 64
UN Guiding Principles, 42–44
UN High Commissioner for Refugees'
  Guidelines on International
  Protection, 12
United Nations Environment Programme
  (UNEP), 19
United Nations Framework Convention on
  Climate Change (UNFCCC), 8
United States Fourth National Climate
  Assessment, 6
Universal Declaration of Human Rights, 45
Unmanned Aerial Vehicles (UAVs), 23, 55, 65
UN Refugee Agency, 25
UN Security Council, 26
Urban areas, 26
Urban renewal and climate-related projects,
  32, 34, 64

V
Violent extremism, 30, 33, 64

W
Warming conditions, 20

X
Xenophobia, 63, 65
  attitudinal prejudices, 25
  discriminatory policies *vs.* foreign
    nationals, 25
  etymology, 10
  nationality and citizenship, 11
    nation-state ideological formations, 10
    populist rhetoric, 10
    rejection of foreigner, 11
    UN Refugee Agency, 25
    xenophobia-induced generalized
      violence, 24
Xenophobia-induced displaced persons, 55, 56

Printed in the United States
By Bookmasters